# Electrical and Electronics Sample Exam

## for the Electrical and Computer PE Exam

## John A. Camara, PE

PPI2PASS.COM

Professional Publications, Inc. • Belmont, California

# Register Your Book at ppi2pass.com/register

- Receive the latest exam news.
- Obtain exclusive exam tips and strategies.
- Receive special discounts.

## Report Errors for This Book

PPI is grateful to every reader who notifies us of a possible error. Your feedback allows us to improve the quality and accuracy of our products. Report errata at **ppi2pass.com/errata**.

**ELECTRICAL AND ELECTRONICS SAMPLE EXAM**
First Edition

Current release of this edition: 6

**Release History**

| date | edition number | revision number | update |
|------|----------------|-----------------|--------|
| Apr 2016 | 1 | 4 | Minor cover updates. |
| Feb 2018 | 1 | 5 | Minor corrections. Minor cover updates. |
| Jul 2018 | 1 | 6 | Minor corrections. Minor formatting and pagination changes. |

PPI
1250 Fifth Avenue, Belmont, CA 94002
(650) 593-9119
ppi2pass.com

Library of Congress Control Number: 2011923328

F E D C B A

# Table of Contents

# Preface and Acknowledgments

The *Electrical and Electronics Sample Exam for the Electrical and Computer PE Exam* is a practice exam designed to the format and specifications defined by the National Council of Examiners for Engineering and Surveying (NCEES) for the Principles and Practice of Engineering (PE) Electrical and Computer—Electronics, Controls, and Communications exam.

The sample exam provides an opportunity for comprehensive exam preparation: the incorrect answers can enlighten; the correct answers are thoroughly explained; and since the problems are based on a range of computer engineering topics, you will recognize areas where further study and preparation are needed. With diligent study, you will find the preparation you seek within these pages. At the very least, the questions will lead you to an intelligent search for information. I hope this book serves you well and that you enjoy the adventure of learning.

Many sources were used to define the scope of this book, including NCEES publications pertaining to the Electrical and Computer—Electronics, Controls, and Communications exam, many references listed in the Codes and References section of this book, and survey comments from recent PE examinees.

Should you find an error in this book, know that it is mine, and that I regret it. Beyond that, I hope two things happen. First, please let me know about the error by using the error reporting form on the PPI website, found at **ppi2pass.com/errata**. Second, I hope you learn something from the error—I know I will! I appreciate suggestions for improvement, additional questions, and recommendations for expansion so that new editions or similar texts will better meet the needs of future examinees.

The *Electrical and Electronics Sample Exam* descended from the third edition of the *Electrical Engineering Sample Examinations.* Thanks go to the very professional and dedicated team at PPI that included Sarah Hubbard, director of product development and implementation, whose guidance from the beginning has been most appreciated; Cathy Schrott, production services manager; Megan Synnestvedt, editorial project manager; Tyler Hayes and Julia White, copy editors; Tom Bergstrom, technical illustrator; Kate Hayes, production associate; and Amy Schwertman, cover designer. And, finally, to Becky Camara, who makes life grand in so many ways.

John A. Camara, PE

# Codes and References

The information that was used to write and update this book was based on the exam specifications at the time of publication. However, as with engineering practice itself, the PE exam is not always based on the most current codes or cutting-edge technology. Similarly, codes, standards, and regulations adopted by state and local agencies often lag issuance by several years. It is likely that the codes that are most current, the codes that you use in practice, and the codes that are the basis of your exam will all be different. However, differences between code editions typically minimally affect the technical accuracy of this book, and the methodology presented remains valid. For more information about the variety of codes related to electrical engineering, refer to the following organizations and their websites.

American National Standards Institute (ansi.org)
Electronic Components Industry Association
(ecianow.org)
Federal Communications Commission (fcc.gov)
Institute of Electrical and Electronics Engineers
(ieee.org)
International Organization for Standardization (iso.org)
International Society of Automation (isa.org)
National Electrical Manufacturers Association
(nema.org)
National Fire Protection Association (nfpa.org)

The PPI website (**ppi2pass.com/faqs**) provides the dates and editions of the codes, standards, and regulations on which NCEES has announced the PE exams are based. It is your responsibility to find out which codes are relevant to your exam.

The minimum recommended library of materials to bring to the Electrical PE exam consists of this book, any applicable code books, a standard handbook of electrical engineering, and one or two textbooks that cover fundamental circuit theory (both electrical and electronic).

## CODES AND STANDARDS

47 CFR 73: *Code of Federal Regulations*, "Title 47—Telecommunication, Part 73—Radio Broadcast Services," 2014. Office of the Federal Register National Archives and Records Administration, Washington, DC. (Communications.)

IEEE/ASTM SI 10: *American National Standard for Metric Practice*, 2010. ASTM International, West Conshohocken, PA. (Metric.)

IEEE Std 141 (IEEE Red Book): *IEEE Recommended Practice for Electric Power Distribution for Industrial Plants*, 1993. The Institute of Electrical and Electronics Engineers, Inc., New York, NY.

IEEE Std 142 (IEEE Green Book): *IEEE Recommended Practice for Grounding of Industrial and Commercial Power Systems*, 2007. The Institute of Electrical and Electronics Engineers, Inc., New York, NY.

IEEE Std 241 (IEEE Gray Book): *IEEE Recommended Practice for Electrical Power Systems in Commercial Buildings*, 1990. The Institute of Electrical and Electronics Engineers, Inc., New York, NY.

IEEE Std 242 (IEEE Buff Book): *IEEE Recommended Practice for Protection and Coordination of Industrial and Commercial Power Systems*, 2001. The Institute of Electrical and Electronics Engineers, Inc., New York, NY.

IEEE Std 399 (IEEE Brown Book): *IEEE Recommended Practice for Industrial and Commercial Power Systems Analysis*, 1997. The Institute of Electrical and Electronics Engineers, Inc., New York, NY.

IEEE Std 446 (IEEE Orange Book): *IEEE Recommended Practice for Emergency and Standby Power Systems for Industrial and Commercial Applications*, 1995. The Institute of Electrical and Electronics Engineers, Inc., New York, NY.

IEEE Std 493 (IEEE Gold Book): *IEEE Recommended Practice for the Design of Reliable Industrial and Commercial Power Systems*, 2007. The Institute of Electrical and Electronics Engineers, Inc., New York, NY.

IEEE Std 551 (IEEE Violet Book): *IEEE Recommended Practice for Calculating Short-Circuit Currents in Industrial and Commercial Power Systems*, 2006. The Institute of Electrical and Electronics Engineers, Inc., New York, NY.

IEEE Std 602 (IEEE White Book): *IEEE Recommended Practice for Electric Systems in Health Care Facilities*, 2007. The Institute of Electrical and Electronics Engineers, Inc., New York, NY.

IEEE Std 739 (IEEE Bronze Book): *IEEE Recommended Practice for Energy Management in Industrial and Commercial Facilities*, 1995. The Institute of Electrical and Electronics Engineers, Inc., New York, NY.

IEEE Std 902 (IEEE Yellow Book): *IEEE Guide for Maintenance, Operation, and Safety of Industrial and Commercial Power Systems*, 1998. The Institute of Electrical and Electronics Engineers, Inc., New York, NY.

IEEE Std 1015 (IEEE Blue Book): *IEEE Recommended Practice for Applying Low-Voltage Circuit Breakers Used in Industrial and Commercial Power Systems*, 2006. The Institute of Electrical and Electronics Engineers, Inc., New York, NY.

IEEE Std 1100 (IEEE Emerald Book): *IEEE Recommended Practice for Powering and Grounding Electronic Equipment*, 2005. The Institute of Electrical and Electronics Engineers, Inc., New York, NY.

NEC (NFPA 70): *National Electrical Code*, 2017. National Fire Protection Association, Quincy, MA. (Power.)

NESC (IEEE C2): *2017 National Electrical Safety Code*, 2016. The Institute of Electrical and Electronics Engineers, Inc., New York, NY. (Power.)

## REFERENCES

Anthony, Michael A. *NEC Answers*. New York, NY: McGraw-Hill. (*National Electrical Code* example applications textbook.)

Bronzino, Joseph D. *The Biomedical Engineering Handbook*. Boca Raton, FL: CRC Press. (Electrical and electronics handbook.)

Chemical Rubber Company. *CRC Standard Mathematical Tables and Formulae*. Boca Raton, FL: CRC Press. (General engineering reference.)

Croft, Terrell and Wilford I. Summers. *American Electricians' Handbook*. New York, NY: McGraw-Hill. (Power handbook.)

Earley, Mark W., et al. *National Electrical Code Handbook*, 2017 ed. Quincy, MA: National Fire Protection Association. (Power handbook.)

Fink, Donald G. and H. Wayne Beaty. *Standard Handbook for Electrical Engineers*. New York, NY: McGraw-Hill. (Power and electrical and electronics handbook.)

Grainger, John J. and William D. Stevenson, Jr. *Power System Analysis*. New York, NY: McGraw-Hill. (Power textbook.)

Horowitz, Stanley H. and Arun G. Phadke. *Power System Relaying*. Chichester, West Sussex: John Wiley & Sons, Ltd. (Power protection textbook.)

Huray, Paul G. *Maxwell's Equations*. Hoboken, NJ: John Wiley & Sons, Inc. (Power and electrical and electronics textbook.)

Jaeger, Richard C. and Travis Blalock. *Microelectronic Circuit Design*. New York, NY: McGraw-Hill Education. (Electronic fundamentals textbook.)

Lee, William C.Y. *Wireless and Cellular Telecommunications*. New York, NY: McGraw-Hill. (Electrical and electronics handbook.)

Marne, David J. *National Electrical Safety Code (NESC) 2017 Handbook*. New York, NY: McGraw-Hill Professional. (Power handbook.)

McMillan, Gregory K. and Douglas Considine. *Process/Industrial Instruments and Controls Handbook*. New York, NY: McGraw-Hill Professional. (Power and electrical and electronics handbook.)

Millman, Jacob and Arvin Grabel. *Microelectronics*. New York, NY: McGraw-Hill. (Electronic fundamentals textbook.)

Mitra, Sanjit K. *An Introduction to Digital and Analog Integrated Circuits and Applications*. New York, NY: Harper & Row. (Digital circuit fundamentals textbook.)

Parker, Sybil P., ed. *McGraw-Hill Dictionary of Scientific and Technical Terms*. New York, NY: McGraw-Hill. (General engineering reference.)

Plonus, Martin A. *Applied Electromagnetics*. New York, NY: McGraw-Hill. (Electromagnetic theory textbook.)

Rea, Mark S., ed. *The IESNA Lighting Handbook: Reference & Applications*. New York, NY: Illuminating Engineering Society of North America. (Power handbook.)

Shackelford, James F. and William Alexander, eds. *CRC Materials Science and Engineering Handbook*. Boca Raton, FL: CRC Press, Inc. (General engineering handbook.)

Van Valkenburg, M.E. and B.K. Kinariwala. *Linear Circuits*. Englewood Cliffs, NJ: Prentice-Hall. (AC/DC fundamentals textbook.)

Wildi, Theodore and Perry R. McNeill. *Electrical Power Technology*. New York, NY: John Wiley & Sons. (Power theory and application textbook.)

# Introduction

## ABOUT THE ELECTRICAL AND COMPUTER—ELECTRONICS, CONTROLS, AND COMMUNICATIONS EXAM

The Electrical and Computer—Electronics, Controls, and Communications exam is made up of 80 problems and is divided into two four-hour long sessions. The format of all exam problems is multiple-choice with a problem statement and all required defining information, followed by four logical choices. Only one of the four options is correct, and the problems are completely independent of each other.

The topics and the distribution of problems for the Electrical and Computer—Electronics, Controls, and Communications exam are as follows. The distribution both on this sample exam and on the NCEES exam is approximate.

**Electrical and Computer—Electronics, Controls, and Communications Exam**

- General Electrical Engineering Knowledge (32 questions):

  circuit analysis; measurement and instrumentation; safety and reliability; signal processing

- Digital Systems (8 questions):

  digital logic; digital components

- Electromagnetics (8 questions):

  electromagnetic fields; guided waves; antennas

- Electronics (16 questions):

  electronics circuits; electronic components and applications

- Control Systems (8 questions):

  analysis and design of analog or digital control systems

- Communications (8 questions):

  modulation techniques; noise and interference; communication systems

According to the NCEES, exam questions related to codes and standards will be based on either (1) an interpretation of a code or standard that is presented in the exam booklet or (2) a code or standard that a committee of licensed engineers feels minimally competent engineers should know.

For further information about the exams, and tips on how to prepare for the exams, consult PPI's website, **ppi2pass.com/faqs**.

## HOW TO USE THIS BOOK

Prior to taking the sample exam in this book, assemble your materials as if you are taking the actual exam. Obtain a copy of any reference or code books you plan to use on the exam. (Use the Codes and References section to determine which supplementary materials you will need for the exam.) Since some states have restrictions on the materials you are allowed to use during the exam, visit **ppi2pass.com/stateboards** to find a link to your state's board of engineering and check for any restrictions. Follow these restrictions when taking the sample exam. Remember that adequate preparation, not an extensive library, is the key to success, both when you take the sample exam and the actual exam.

After you feel that you have sufficiently prepared for the sample exam in this book, read the instructions at the beginning of the exam for guidance on how to properly simulate the exam. Set a timer for four hours, and take the morning session. After a one-hour break, turn to the afternoon exam, set the timer, and complete the afternoon session.

Next, check your answers and read through the solutions of the problems that you answered incorrectly or were unable to answer. Evaluate your strengths and weaknesses and select additional texts to supplement your studies. Check the PPI website at **ppi2pass.com** for the latest in exam preparation materials.

The keys to exam success are knowing the basics and practicing solving as many problems as possible. This book will assist you with both objectives.

# Morning Session Instructions

In accordance with the rules established by your state, you may use textbooks, handbooks, bound reference materials, and any approved battery- or solar-powered, silent calculator to work this examination. However, no blank papers, writing tablets, unbound scratch paper, or loose notes are permitted. Sufficient room for scratch work is provided in the Examination Booklet.

You are not permitted to share or exchange materials with other examinees. However, the books and other resources used in this morning session may be changed prior to the afternoon session.

You will have four hours in which to work this session of the examination. Your score will be determined by the number of questions that you answer correctly. There is a total of 40 questions. All 40 questions must be worked correctly in order to receive full credit on the exam. There are no optional questions. Each question is worth 1 point. The maximum possible score for this section of the examination is 40 points.

Partial credit is not available. No credit will be given for methodology, assumptions, or work written in your Examination Booklet.

Record all of your answers on the Answer Sheet. No credit will be given for answers marked in the Examination Booklet. Mark your answers with a no. 2 pencil. Answers marked in pen may not be graded correctly. Marks must be dark and must completely fill the bubbles. Record only one answer per question. If you mark more than one answer, you will not receive credit for the question. If you change an answer, be sure the old bubble is erased completely; incomplete erasures may be misinterpreted as answers.

If you finish early, check your work and make sure that you have followed all instructions. After checking your answers, you may turn in your Examination Booklet and Answer Sheet and leave the examination room. Once you leave, you will not be permitted to return to work or change your answers.

When permission has been given by your proctor, break the seal on the Examination Booklet. Check that all pages are present and legible. If any part of your Examination Booklet is missing, your proctor will issue you a new Booklet.

Do not work any questions from the Afternoon Session during the first four hours of this exam.

WAIT FOR PERMISSION TO BEGIN

Name: _____
       Last           First         Middle Initial

Examinee number: _____

Examination Booklet number: _____

## Principles and Practice of Engineering Examination

## Morning Session Sample Examination

1. Ⓐ Ⓑ Ⓒ Ⓓ       11. Ⓐ Ⓑ Ⓒ Ⓓ       21. Ⓐ Ⓑ Ⓒ Ⓓ       31. Ⓐ Ⓑ Ⓒ Ⓓ
2. Ⓐ Ⓑ Ⓒ Ⓓ       12. Ⓐ Ⓑ Ⓒ Ⓓ       22. Ⓐ Ⓑ Ⓒ Ⓓ       32. Ⓐ Ⓑ Ⓒ Ⓓ
3. Ⓐ Ⓑ Ⓒ Ⓓ       13. Ⓐ Ⓑ Ⓒ Ⓓ       23. Ⓐ Ⓑ Ⓒ Ⓓ       33. Ⓐ Ⓑ Ⓒ Ⓓ
4. Ⓐ Ⓑ Ⓒ Ⓓ       14. Ⓐ Ⓑ Ⓒ Ⓓ       24. Ⓐ Ⓑ Ⓒ Ⓓ       34. Ⓐ Ⓑ Ⓒ Ⓓ
5. Ⓐ Ⓑ Ⓒ Ⓓ       15. Ⓐ Ⓑ Ⓒ Ⓓ       25. Ⓐ Ⓑ Ⓒ Ⓓ       35. Ⓐ Ⓑ Ⓒ Ⓓ
6. Ⓐ Ⓑ Ⓒ Ⓓ       16. Ⓐ Ⓑ Ⓒ Ⓓ       26. Ⓐ Ⓑ Ⓒ Ⓓ       36. Ⓐ Ⓑ Ⓒ Ⓓ
7. Ⓐ Ⓑ Ⓒ Ⓓ       17. Ⓐ Ⓑ Ⓒ Ⓓ       27. Ⓐ Ⓑ Ⓒ Ⓓ       37. Ⓐ Ⓑ Ⓒ Ⓓ
8. Ⓐ Ⓑ Ⓒ Ⓓ       18. Ⓐ Ⓑ Ⓒ Ⓓ       28. Ⓐ Ⓑ Ⓒ Ⓓ       38. Ⓐ Ⓑ Ⓒ Ⓓ
9. Ⓐ Ⓑ Ⓒ Ⓓ       19. Ⓐ Ⓑ Ⓒ Ⓓ       29. Ⓐ Ⓑ Ⓒ Ⓓ       39. Ⓐ Ⓑ Ⓒ Ⓓ
10. Ⓐ Ⓑ Ⓒ Ⓓ      20. Ⓐ Ⓑ Ⓒ Ⓓ       30. Ⓐ Ⓑ Ⓒ Ⓓ       40. Ⓐ Ⓑ Ⓒ Ⓓ

# Morning Session Exam

**1.** A magnetic force vector is represented by $3\mathbf{i} + 6\mathbf{j} + 9\mathbf{k}$. Of the following, which vector has the same direction as the magnetic force vector and possesses a magnitude of one?

(A)  $0.020\mathbf{i} + 0.050\mathbf{j} + 0.070\mathbf{k}$

(B)  $0.27\mathbf{i} + 0.53\mathbf{j} + 0.80\mathbf{k}$

(C)  $0.71\mathbf{i} + 1.4\mathbf{j} + 2.1\mathbf{k}$

(D)  $1.0\mathbf{i} + 2.0\mathbf{j} + 3.0\mathbf{k}$

**2.** An electrical experiment is conducted and the results are plotted. Using regression techniques, the equation describing the graph is determined to be $y(x) = \sin x - e^x$. Using differential calculus, determine the coordinates of the critical point in the range $[-1, 10]$, and determine whether the point is a local maximum or minimum.

(A)  $(0, -1)$, minimum

(B)  $(0, -1)$, maximum

(C)  $(1.5, 3.5)$, minimum

(D)  $(1.5, 3.5)$, maximum

**3.** What is the gradient of the three-dimensional function?

$$f(x,\ y,\ z) = 4x^2 - xy + 3z$$

(A)  $8x\mathbf{i} + x\mathbf{j} + 3\mathbf{k}$

(B)  $8x\mathbf{i} - y\mathbf{j} + 3\mathbf{k}$

(C)  $8x\mathbf{i} + xy\mathbf{j} + 3\mathbf{k}$

(D)  $(8x - y)\mathbf{i} - x\mathbf{j} + 3\mathbf{k}$

**4.** What time-domain function is associated with the given Laplace transform?

$$F(s) = \frac{2}{s^3(s - 3)}$$

(A)  $t^2 e^3$

(B)  $t^2(1 - e^{3t})$

(C)  $-\frac{2}{9}e^{3t} + t^2 + \frac{t}{3} + \frac{2}{9}$

(D)  $\frac{2}{27}e^{3t} - \frac{1}{3}t^2 - \frac{2}{9}t - \frac{2}{27}$

**5.** For the circuit shown, the input voltage, $v_{\text{in}}$, is $(1\ \text{V})\sin 377t$. The output voltage is $0$ V at $t = 0$ s.

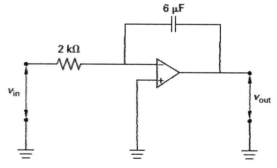

Assuming ideal conditions, what is the expression for the output voltage of the circuit when $t \geq 0$ s?

(A)  $(0.221\ \text{V})(\cos(377t - 1))$

(B)  $(0.221\ \text{V})\cos 377t$

(C)  $(1.13\ \text{V})\sin 377t$

(D)  $(1.13\ \text{V})(\sin(377t + 1))$

**6.** For the circuit shown, what is the voltage at node A?

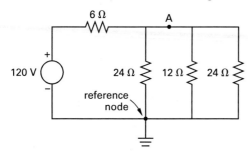

(A)  12 V

(B)  30 V

(C)  60 V

(D)  96 V

**7.** Which of the following amplifier circuits, designed using a single op amp, would exhibit a 0.75 gain, an 800 kΩ input resistance, and a 0 Ω output resistance?

(A)

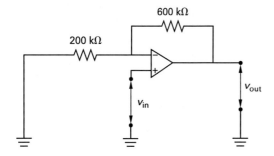

(B)

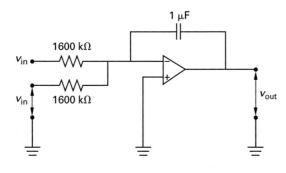

(C)

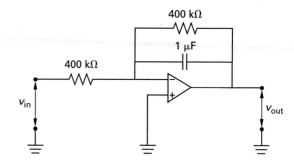

(D)

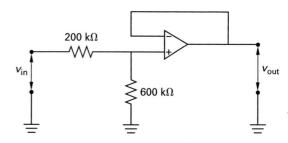

**8.** The given tuning circuit is used in a communication circuit and the internal resistance of the inductor is 3 Ω.

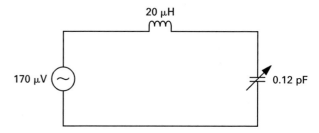

What is most nearly the resonant frequency of the circuit?

(A)  0.15 MHz

(B)  16 MHz

(C)  100 MHz

(D)  250 MHz

**9.** A communications line uses a shield twisted pair (STP) cable that has a velocity factor of 0.6. What is most nearly the wavelength of a 1 GHz signal on the STP cable?

(A)  0.2 m

(B)  0.3 m

(C)  0.6 m

(D)  1 m

**10.** A communication circuit senses the following voltage waveform.

$$v(t) = 0.5 \text{ V} + (1.7 \text{ V})\sin(2513t + \theta)$$

The phase angle, $\theta$, is 2 rad. $t$ is measured in seconds. What is most nearly the period of the waveform?

(A)  1.2 ms

(B)  2.5 ms

(C)  25 ms

(D)  400 ms

**11.** A given function is represented as

$$3e^{-2t}\cos(377t - 30°)$$

Assume that the cosine term will be used as the reference. The periodic portion of the function will be represented by $e^{st}$. What is the value of **s**?

(A)  $-2 + j3$

(B)  $\dfrac{\pi}{6} + j3$

(C)  $3 + j\left(\dfrac{\pi}{6}\right)$

(D)  $-2 + j377$

**12.** A portion of an electronic sensing circuit follows.

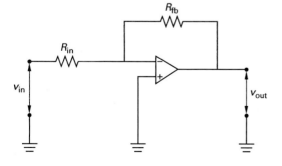

Which of the following correctly expresses the output voltage, $v_{out}$, of this circuit?

(A)  $v_{in}$

(B)  $v_{in}\left(-\dfrac{R_{fb}}{R_{in}}\right)$

(C)  $v_{in}\left(1 + \dfrac{R_{fb}}{R_{in}}\right)$

(D)  $\left(\dfrac{R_{fb}}{R_{in}}\right)(v^+ - v^-)$

**13.** In the parallel analog-to-digital converter shown, the op amp supply voltages are $\pm 5$ V, and the reference voltage, $V_{ref}$, is 1 V.

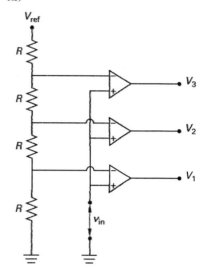

The output voltage is considered high at +5 V and low at −5 V. If $v_{in} = 0.76$ V, what is the output, assuming an output order of $V_3$, $V_2$, $V_1$?

(A)  L, L, L

(B)  L, L, H

(C)  L, H, H

(D)  H, H, H

**14.** For the following op amp configuration, used to provide gain in an electronic circuit, what is the expression for $v_{out}$?

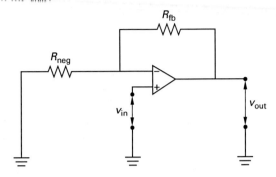

(A) $v_{in}$

(B) $v_{in}\left(-\dfrac{R_{fb}}{R_{neg}}\right)$

(C) $v_{in}\left(1+\dfrac{R_{fb}}{R_{neg}}\right)$

(D) $\left(\dfrac{R_{fb}}{R_{neg}}\right)(v^+ - v^-)$

**15.** For the circuit shown, which is being used as a buffer, what is the expression for the output voltage?

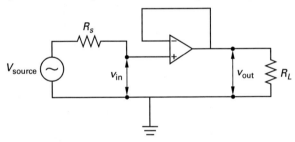

(A) $v_{in}$

(B) $-R_s v_{in}$

(C) $v_{in}(1+R_s)$

(D) $R_s(v^+ - v^-)$

**16.** The current through a 90 mH inductor is given by

$$i(t) = (8\ \text{A})\sin 25t$$

What is the voltage at $t = \pi/50$ s?

(A) 0.0 V

(B) 18 V

(C) 20 V

(D) 200 V

**17.** What is the magnitude of the current through the 18 Ω resistor shown?

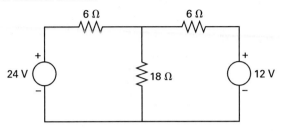

(A) 0 A

(B) 4/25 A

(C) 1/2 A

(D) 6/7 A

**18.** In the simplified circuit shown, approximately how much time is required for any given transient within the circuit to reach steady state?

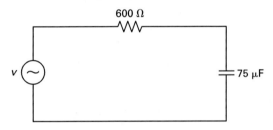

(A) 0.020 s

(B) 0.045 s

(C) 0.080 s

(D) 0.22 s

**19.** In the control circuit shown, what is the value of the current, $I$, in amperes?

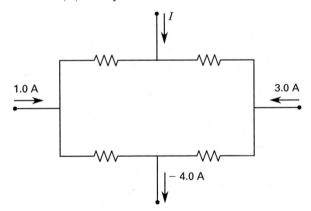

(A) −8.0 A

(B) 0.0 A

(C) 1.0 A

(D) 8.0 A

**20.** A capacitance meter reads 15 $\mu$F when placed across the terminals of an electrolytic capacitor. A voltmeter across the same terminals reads 90 V. What is most nearly the charge on the capacitor?

(A)   0.14 $\mu$C

(B)   14 $\mu$C

(C)   1.4 mC

(D)   6.0 mC

**21.** What is the magnitude of the Norton equivalent current for the circuit shown, with $R = 2\ \Omega$?

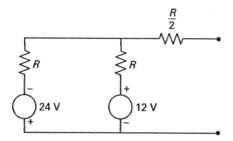

(A)   3.0 A

(B)   6.0 A

(C)   9.0 A

(D)   12 A

**22.** Regarding the Fourier series of the waveform shown, which of the following statements is FALSE?

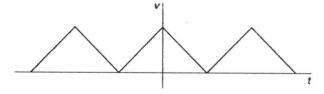

(A)   The series has even symmetry.

(B)   The series contains cosines.

(C)   The first term of the series is 0.

(D)   The $b_n$ coefficients are 0.

**23.** A control system uses the following input filter circuitry.

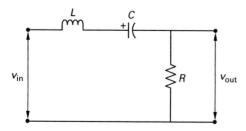

The input impedance normalized to $R$ (i.e., $|\mathbf{Z}_{in}/R|$) for various input filter circuitry is shown. Which is the actual frequency response of the input filter circuitry?

(A)

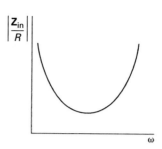

(B)

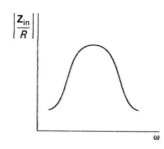

(C)

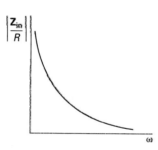

(D)

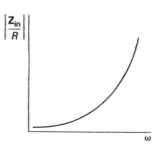

**24.** The RC circuit shown contains a capacitor with an initial voltage of $v_C(0)$.

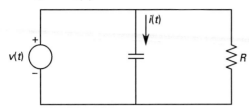

What is the expression for the current $I(s)$?

(A) $sCV(s) - sCv_C(s)$

(B) $C\big(sV(s) - v_C(0)\big)$

(C) $\dfrac{R}{\dfrac{1}{sC} + R}$

(D) $\dfrac{V(s)}{sC - v_C(s)}$

**25.** Which of the following types of electric charge motion generates a steadily radiating (i.e., traveling) electromagnetic wave?

(A) accelerating

(B) displacement

(C) static

(D) uniform

**26.** Magnetic flux density, **B**, and magnetic field strength, **H**, may experience changes at the interface of materials whose magnetic properties differ from one another. Consider the following.

  i. magnitude of **B**

  ii. normal component of **B**

  iii. tangential component of **B**

  iv. magnitude of **H**

  v. normal component of **H**

  vi. tangential component of **H**

What combination represents the properties of an electromagnetic wave that are continuous (i.e., do not change) across such an interface?

(A) i and ii

(B) ii and vi

(C) iii and v

(D) iv and vi

**27.** What is the intrinsic impedance, $\eta_0$, of free space?

(A) $3.00 \times 10^{-3}\ \Omega$

(B) $2.65\ \Omega$

(C) $377\ \Omega$

(D) $1.42 \times 10^5\ \Omega$

**28.** The **E** field within a material is given by

$$\mathbf{E} = \left(175\ \frac{\text{V}}{\text{m}}\right)\sin 10^9 t$$

The relative permittivity, $\epsilon_r$, is 1. The direction of the field is constant. What is the expression for the magnitude of the displacement current density?

(A) $\left(1.55\ \dfrac{\text{A}}{\text{m}^2}\right)\cos 10^9 t$

(B) $\left(1.55\ \dfrac{\text{A}}{\text{m}^2}\right)\sin 10^9 t$

(C) $\left(1.75 \times 10^2\ \dfrac{\text{A}}{\text{m}^2}\right)\cos 10^9 t$

(D) $\left(1.75 \times 10^{11}\ \dfrac{\text{A}}{\text{m}^2}\right)\cos 10^9 t$

**29.** Which of the following four mathematical statements related to Maxwell's equations is FALSE?

(A) $\nabla \cdot \mathbf{D} = 0$

(B) $\nabla \cdot \mathbf{D} = \rho$

(C) $\nabla \cdot \mathbf{B} = 0$

(D) $\nabla \cdot \mathbf{B} = \rho$

**30.** An amplifier consists of multiple stages with gains as shown. A voltage-divider network provides the feedback.

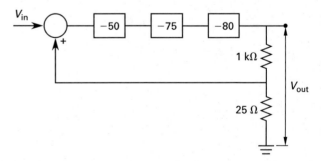

What is most nearly the overall gain of the network, with feedback?

(A)  $-300\,000$

(B)  $-40$

(C)  $0.03$

(D)  $40$

**31.** The transfer function of a given filtering network is

$$T(s) = \frac{-10^3}{s^2 + 20s + 10^6}$$

What is the quality factor of the filter?

(A)  $20$

(B)  $50$

(C)  $50\,000$

(D)  $1\,000\,000$

**32.** A control system is modeled as a five-block system with two summing points as shown.

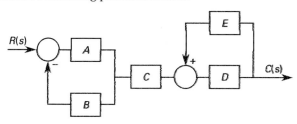

What is the overall transfer function of the control system?

(A)  $C$

(B)  $\dfrac{ACD}{1 - BE}$

(C)  $\left(\dfrac{A}{1 - AB}\right)\left(\dfrac{CD}{1 + ED}\right)$

(D)  $\dfrac{CAD}{(1 + AB)(1 - ED)}$

**33.** A system is acted upon by a constant force of two units commencing at $t = 0$ s. The transfer function is shown. What is the time-based response function, $r(t)$?

$$T(s) = \frac{6}{s + 2}$$

(A)  $12e^{-12t}$

(B)  $6e^{2t}$

(C)  $1 - e^{-2t}$

(D)  $6 - 6e^{-2t}$

**34.** A response function is given by

$$R(s) = \frac{1}{s(s + 1)}$$

What is the value of the response function $r(t)$ as $t$ approaches $\infty$; that is, what is the final steady-state value?

(A)  $0$

(B)  $1$

(C)  $\infty$

(D)  indeterminate

**35.** Which of the following statements referring to a controller effect known as quantization is FALSE?

(A)  Quantization forces a minimum nonzero change in a measurement for any sampled calculation.

(B)  Quantization could also be called resolution.

(C)  The control system challenge is an effective derivative action while keeping quantization within the bounds of the accuracy of the data.

(D)  The use of a "bucket brigade" algorithm approximates the effects of quantization.

**36.** The transfer function for a given network is

$$T(s) = \frac{3(s+4)}{(s+2)(s^2+2s+2)}$$

Which of the following illustrations represents the pole-zero diagram for the given transfer function?

(A)

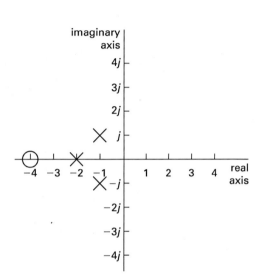

(B)

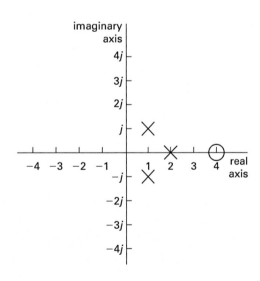

(C)

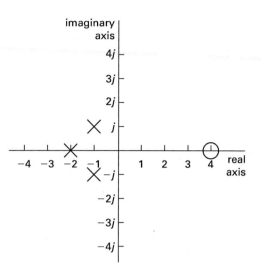

(D)

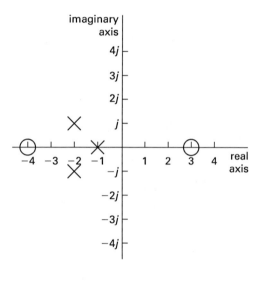

**37.** Consider the following generic Bode plots of gain and phase angle.

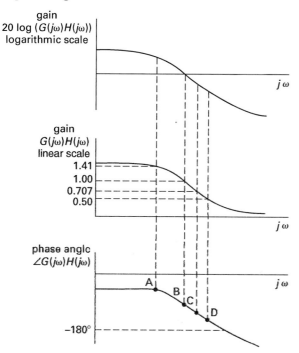

What point represents the phase margin?

(A)   point A

(B)   point B

(C)   point C

(D)   point D

**38.** For a transmission line with a characteristic impedance of 60 $\Omega$ and a terminating resistance of 120 $\Omega$, what is the reflection coefficient?

(A)   0.33

(B)   0.50

(C)   2.0

(D)   3.0

**39.** A space-based antenna transmits at 2 W with a gain of 17 dBW. The path loss is 200 dBW. The receiving antenna has a gain of 50. What is most nearly the effective isotropically radiated power (EIRP) in dBW?

(A)   −170 dBW

(B)   −110 dBW

(C)   15 dBW

(D)   20 dBW

**40.** For a modulated FM signal with the following form, what is the approximate maximum frequency deviation from the carrier frequency?

$$s_{\text{mod}}(t) = A\cos(5.5 \times 10^{11} t + 8\sin 10^{6} t)$$

(A)   $8.8 \times 10^5$ Hz

(B)   $1.0 \times 10^6$ Hz

(C)   $1.3 \times 10^6$ Hz

(D)   $8.0 \times 10^6$ Hz

# STOP!

## DO NOT CONTINUE!

This concludes the Morning Session of the examination. If you finish early, check your work and make sure that you have followed all instructions. After checking your answers, you may turn in your examination booklet and answer sheet and leave the examination room. Once you leave, you will not be permitted to return to work or change your answers.

# Afternoon Session Instructions

In accordance with the rules established by your state, you may use textbooks, handbooks, bound reference materials, and any approved battery- or solar-powered, silent calculator to work this examination. However, no blank papers, writing tablets, unbound scratch paper, or loose notes are permitted. Sufficient room for scratch work is provided in the Examination Booklet.

You are not permitted to share or exchange materials with other examinees. However, the books and other resources used in this afternoon session do not have to be the same as were used in the morning session.

You will have four hours in which to work this session of the examination. Your score will be determined by the number of questions that you answer correctly. There is a total of 40 questions. All 40 questions must be worked correctly in order to receive full credit on the exam. There are no optional questions. Each question is worth 1 point. The maximum possible score for this section of the examination is 40 points.

Partial credit is not available. No credit will be given for methodology, assumptions, or work written in your Examination Booklet.

Record all of your answers on the Answer Sheet. No credit will be given for answers marked in the Examination Booklet. Mark your answers with a no. 2 pencil. Answers marked in pen may not be graded correctly. Marks must be dark and must completely fill the bubbles. Record only one answer per question. If you mark more than one answer, you will not receive credit for the question. If you change an answer, be sure the old bubble is erased completely; incomplete erasures may be misinterpreted as answers.

If you finish early, check your work and make sure that you have followed all instructions. After checking your answers, you may turn in your Examination Booklet and Answer Sheet and leave the examination room. Once you leave, you will not be permitted to return to work or change your answers.

When permission has been given by your proctor, break the seal on the Examination Booklet. Check that all pages are present and legible. If any part of your Examination Booklet is missing, your proctor will issue you a new Booklet.

Do not work any questions from the Morning Session during the second four hours of this exam.

WAIT FOR PERMISSION TO BEGIN

Name: _____
      Last            First           Middle Initial

Examinee number: _____

Examination Booklet number: _____

## Principles and Practice of Engineering Examination

## Afternoon Session Sample Examination

41. (A) (B) (C) (D)      51. (A) (B) (C) (D)      61. (A) (B) (C) (D)      71. (A) (B) (C) (D)
42. (A) (B) (C) (D)      52. (A) (B) (C) (D)      62. (A) (B) (C) (D)      72. (A) (B) (C) (D)
43. (A) (B) (C) (D)      53. (A) (B) (C) (D)      63. (A) (B) (C) (D)      73. (A) (B) (C) (D)
44. (A) (B) (C) (D)      54. (A) (B) (C) (D)      64. (A) (B) (C) (D)      74. (A) (B) (C) (D)
45. (A) (B) (C) (D)      55. (A) (B) (C) (D)      65. (A) (B) (C) (D)      75. (A) (B) (C) (D)
46. (A) (B) (C) (D)      56. (A) (B) (C) (D)      66. (A) (B) (C) (D)      76. (A) (B) (C) (D)
47. (A) (B) (C) (D)      57. (A) (B) (C) (D)      67. (A) (B) (C) (D)      77. (A) (B) (C) (D)
48. (A) (B) (C) (D)      58. (A) (B) (C) (D)      68. (A) (B) (C) (D)      78. (A) (B) (C) (D)
49. (A) (B) (C) (D)      59. (A) (B) (C) (D)      69. (A) (B) (C) (D)      79. (A) (B) (C) (D)
50. (A) (B) (C) (D)      60. (A) (B) (C) (D)      70. (A) (B) (C) (D)      80. (A) (B) (C) (D)

# Afternoon Session Exam

**41.** Consider the electronic circuit models shown. Identifying subscripts have been removed. Which of these models represents a field-effect transistor?

(A)

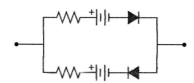

(B)

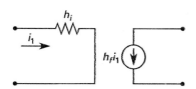

(C)

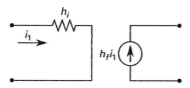

(D)

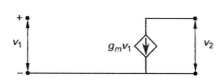

**42.** The force acting on a particle is determined to be $2\mathbf{i} + 3\mathbf{j} + \mathbf{k}$. The particle is displaced from point A to point B along a straight line segment, AB. The $(x, y, z)$ coordinates of point A are $(0, 3, 1)$. The $(x, y, z)$ coordinates of point B are $(-1, 3, -1)$. Units for the force and distance are consistent. What is the value of the work done on the particle?

(A) $-4$

(B) $0$

(C) $2$

(D) $\sqrt{10}$

**43.** The rectangular form of a given complex number is

$$3 + 4j$$

What is most nearly the number when using trigonometric functions?

(A) $5e^{j36.86°}$

(B) $(5)(\cos 36.86° + j\sin 36.86°)$

(C) $\cos 0.64 + j\sin 0.64$

(D) $(5)(\cos 0.93 + j\sin 0.93)$

**44.** An electronic voltmeter using a d'Arsonval meter display is shown. The full-scale meter current is 0.1 mA. The voltmeter reads 250 $V_{rms}$. To ensure linear operation, the op amp input and output are restrained to operate a minimum of 3 V from the rail voltages (power supply voltages).

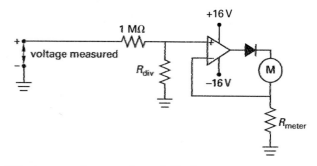

What approximate value of divider resistance, $R_{div}$, is necessary to ensure linear operation?

(A) 40 k$\Omega$

(B) 50 k$\Omega$

(C) 55 k$\Omega$

(D) 1000 k$\Omega$

**45.** An inductance of 5 mH has a current given by $i(t) = 1.5 \text{ A}(1 - e^{-1000t})$. What is most nearly the value of the voltage 100 $\mu$s into a transient?

(A) 0.0 V

(B) 6.8 V

(C) 7.5 V

(D) 13 V

**46.** A point charge of $-1.6 \times 10^{19}$ C is moved along the $x$-axis (units of meters) from $(0, 0, 0)$ to $(4, 0, 0)$ in an electrostatic field given by

$$\mathbf{E} = (2x + 4y)\mathbf{i} + 2x\,\mathbf{j}$$

What is most nearly the work done on the charge?

(A) $-12.8 \times 10^{19}$ J

(B) $-3.20 \times 10^{19}$ J

(C) $1.60 \times 10^{19}$ J

(D) $2.56 \times 10^{20}$ J

**47.** Given a value of $\mathbf{D} = 1.5 \times 10^{-6}\mathbf{a}_r$ C/m$^2$ and a relative permittivity of 2.9, what is the approximate electric field strength?

(A) $3.8 \times 10^{-17}\mathbf{a}_r$ N/C

(B) $5.2 \times 10^{-7}\mathbf{a}_r$ N/C

(C) $5.8 \times 10^4\mathbf{a}_r$ N/C

(D) $4.4 \times 10^6\mathbf{a}_r$ N/C

**48.** The voltage in a circuit is represented by the complex number $3 + j5$ V. The current is represented by the complex number $4 - j10$ A. Which complex number represents the impedance?

(A) $0.10 + j0.43\ \Omega$

(B) $-0.47 + j0.20\ \Omega$

(C) $-0.33 + j0.43\ \Omega$

(D) $0.14 + j0.60\ \Omega$

**49.** Consider the following figure of a MOSFET.

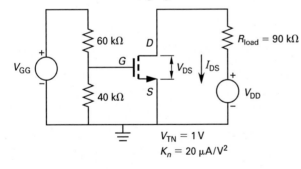

What is the gate-source voltage, $V_{GS}$, for the indicated configuration if the gate supply voltage, $V_{GG}$, is 10.0 V?

(A) 0 V

(B) 0.3 V

(C) 0.7 V

(D) 4.0 V

**50.** An optical transmission line is known to have an absolute index of refraction, $n$, of 2. Air generally approximates free space conditions for waves. What angle of incidence is required to obtain total internal reflection?

(A) $\pi/6$ rad

(B) $\pi/4$ rad

(C) $\pi/3$ rad

(D) $\pi/2$ rad

**51.** A radio station transmits 100 kW with a directivity gain of 1.5. What is the approximate maximum radiated power density 5 km from the antenna?

(A) 0.5 mW/m$^2$

(B) 3.0 mW/m$^2$

(C) 2.5 W/m$^2$

(D) 6.0 W/m$^2$

**52.** Consider the following circuit.

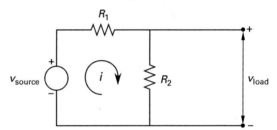

What is the open-loop transfer function for $v_{\text{load}}$?

(A) $v_{\text{load}} = iR_2$

(B) $i = \dfrac{v_{\text{source}}}{R_1 + R_2}$

(C) $v_{\text{load}} = \left(\dfrac{R_2}{R_1 + R_2}\right)v_{\text{source}}$

(D) $v_{\text{load}} = \left(\dfrac{v_{\text{source}} - v_{\text{load}}}{R_1}\right)R_2$

**53.** The canonical form of a typical feedback control system is shown in the following figure.

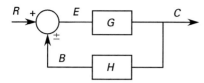

What expression represents the open-loop transfer function?

(A)  $B/R$

(B)  $C/R$

(C)  $E/H$

(D)  $GH$

**54.** A silicon transistor with coupling capacitors sized to operate as short circuits at the frequency of interest is shown.

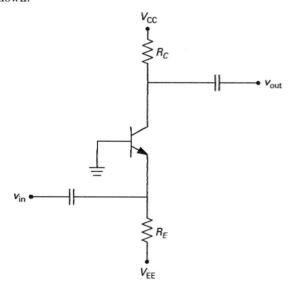

What type of configuration is shown and what type of property is associated with such a configuration?

(A)  common emitter, high $A_V$

(B)  common emitter, high $R_i$

(C)  common base, high $A_V$

(D)  common base, high $R_i$

**55.** In the amplifier circuitry shown, the transistor utilizes silicon and the coupling capacitors are sized to act as short circuits at the frequency of interest.

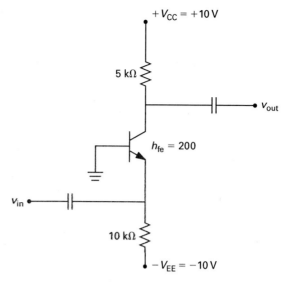

If the input voltage signal is sinusoidal, what approximate magnitude of input signal saturates the transistor, ending its amplification function?

(A)  2.4 V

(B)  2.6 V

(C)  4.8 V

(D)  5.4 V

**56.** A given transistor circuit uses voltage-divider bias as shown.

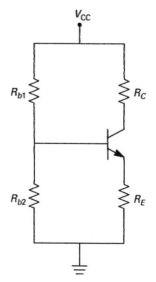

What is the primary purpose of the emitter resistor, $R_E$?

(A)  It prevents thermal runaway.

(B)  It increases sensitivity.

(C)  It decreases output resistance.

(D)  It sets the $Q$-point.

**57.** The closed-loop characteristic polynomial for a control system is given by $s^2 + 4s + 144$. What is most nearly the system damping ratio?

(A)  0.010

(B)  0.17

(C)  4.0

(D)  12

**58.** Which of the following statements referring to commercial AM radio broadcasts in the United States is FALSE?

(A)  The carrier signal is suppressed.

(B)  The modulated signal is frequency shifted.

(C)  Both sidebands are utilized.

(D)  The signal is generally DSB-LC.

**59.** A system has the following characteristic equation.

$$s^3 + 6s^2 + 3s + C = 0$$

Which of the following values of $C$ ensures system stability?

(A)  $-6$

(B)  $-5$

(C)  11

(D)  30

**60.** A digital voltmeter using a 20 k$\Omega$ resistor is designed to measure over a frequency range of 0 Hz to 1000 Hz. What is most nearly the expected thermal agitation noise, or Johnson noise, at 25°C?

(A)  0.03 pV

(B)  0.3 pV

(C)  0.2 $\mu$V

(D)  0.6 $\mu$V

**61.** The magnetic field strength (in A/m) in a region of space is given by

$$\mathbf{H} = (5x + 3y)\mathbf{i} - (5y + 3z)\mathbf{j} + (5z - 3x)\mathbf{k}$$

The curl of the magnetic field strength determines the current density, $\mathbf{J}$ (in A/m$^2$). What is $\mathbf{J}$ in this area of space?

(A)  $3\mathbf{i} + 3\mathbf{j} - 3\mathbf{k}$ A/m$^2$

(B)  $0\mathbf{i} + 3\mathbf{j} - 3\mathbf{k}$ A/m$^2$

(C)  $-3\mathbf{i} + 3\mathbf{j} + 0\mathbf{k}$ A/m$^2$

(D)  $-3\mathbf{i} - 3\mathbf{j} + 6\mathbf{k}$ A/m$^2$

**62.** The RF circuit shown uses a coupler that results in meter power, $P_m$, that is 3 dB less than the input coupler power, $P_{c,in}$.

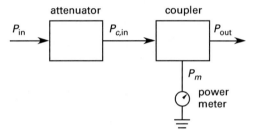

The attenuator results in a 10 dB loss over the frequency range of operation. The input power available is 30 mW. Determine the reading (in mW) on the power meter.

(A)  1.50 mW

(B)  1.77 mW

(C)  14.7 mW

(D)  17.0 mW

**63.** A digital voltmeter that is an integral part of a communications monitoring circuit is designed to operate under the following conditions.

$$T_{\text{ambient}} = 25°C$$
$$\text{BW} = 100 \text{ kHz}$$
$$P_{\text{in}} = 2 \text{ mW}$$

Determine the approximate thermal agitation noise at ambient conditions generated by a 10 k$\Omega$ resistor within the voltmeter.

(A)  1 pV

(B)  1 $\mu$V

(C)  4 $\mu$V

(D)  4 mV

**64.** The circuitry for an electronic voltmeter using a d'Arsonval display is shown.

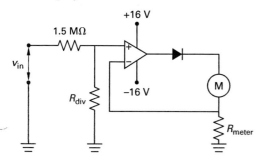

The full-scale meter current of 0.1 mA provides an rms reading of 200 V. Linear operation is ensured by maintaining the op amp input and output to within 3 V of the power-supply voltages. Determine the approximate value of divider resistance, $R_{div}$, necessary to ensure linear operation, assuming a sinusoidal input.

(A)   0.01 MΩ

(B)   0.07 MΩ

(C)   0.2 MΩ

(D)   0.6 MΩ

**65.** The input, $v_{in}$, to the computer circuitry shown is adjusted to the manufacturer's specification for obtaining a logic 1 output.

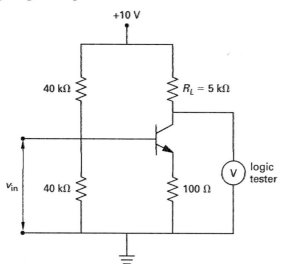

The data sheet provided with the circuitry indicates that the logic 1 condition occurs when the base-emitter junction and the base-collector junction are forward biased. An instrument called a logic tester is attached to the output terminals. The logic tester has two possible

settings: positive logic and negative logic. Determine the approximate current through the load resistor and the position of the logic tester switch.

(A)   0 mA, negative

(B)   0 mA, positive

(C)   2 mA, negative

(D)   2 mA, positive

**66.** A portion of a control system circuit used as an on-off switch for the frequency of the incoming signal is shown.

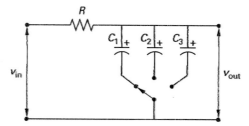

For the particular model of filter installed in the control system, the resistance is 50 kΩ. The values of the selectable capacitances are 0.01 μF, 0.05 μF, and 0.1 μF for $C_1$, $C_2$, and $C_3$, respectively. The "3 dB downpoint" is used as the cutoff. A control system is meant to respond to signals from the DC level to 60 Hz AC. Which capacitor meets this requirement?

(A)   $C_1$

(B)   $C_2$

(C)   $C_3$

(D)   none of the above

**67.** The following circuit is used to measure the difference in voltage between inputs $V_1$ and $V_2$. The op amp properties of this circuit closely approximate those of an ideal op amp.

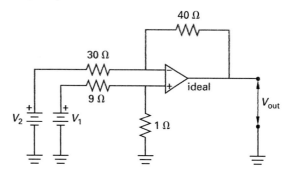

If $V_1 = 11$ V and $V_2 = 12$ V, what is the approximate output voltage?

(A)  −16 V

(B)  −13 V

(C)  1.0 V

(D)  15 V

**68.** A network-powered broadband communications system requires grounding. What is the minimum size of the grounding conductor?

(A)  insulated AWG 6

(B)  insulated AWG 14

(C)  uninsulated AWG 6

(D)  uninsulated AWG 14

**69.** For the battery charging circuit shown, the circuit source voltage is given by $V_{charge} = (170 \text{ V})\sin 377t$.

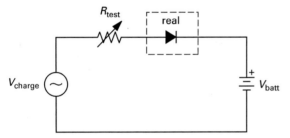

The test procedure requires that at the beginning of the charge, the charge current should not exceed 4 A$_{rms}$. The maximum difference between the battery voltage and the charge voltage at the battery terminals is 5 V$_{rms}$. The battery voltage at the beginning of a charge is expected to be 30 V. The silicon diode can be analyzed using the constant voltage drop model. What is the minimum value of the test resistor that meets all requirements?

(A)  12 Ω

(B)  13 Ω

(C)  31 Ω

(D)  33 Ω

**70.** An AM radio station advertises their frequency as 680 kHz. What frequency is the local oscillator of an AM broadcast receiver set to in order to receive the advertised station?

(A)  225 kHz

(B)  455 kHz

(C)  680 kHz

(D)  1135 kHz

**71.** The FCC Rules, that is, the chapters associated with the requirements in the Code of Federal Regulations, limit a radio station to 5 kW power to avoid interference. The directivity gain of the antenna used is 1.4. What is most nearly the power density of the signal 5 km from the transmitting antenna?

(A)  22 $\mu$W/m$^2$

(B)  0.10 mW/m$^2$

(C)  20 W/m$^2$

(D)  2.0 kW/m$^2$

**72.** A typical residential unit uses AWG 12 wire containing approximately $2.77 \times 10^{23}$ free electrons per meter of wire. A typical household breaker is rated at 15 A. In 1 min, approximately what percentage of the free electrons pass through a breaker carrying rated current?

(A)  0.08%

(B)  0.1%

(C)  2%

(D)  5%

**73.** A given circuit element is represented by the following current and voltage equations, where $t$ is in seconds. What is most nearly the total energy transferred from $t \geq 0$ s?

$$i(t) = (5 \text{ A})e^{-3000t}$$
$$v(t) = (25 \text{ V})(1 - e^{-3000t})$$

(A)  0 mJ

(B)  20 mJ

(C)  30 mJ

(D)  60 mJ

**74.** An electric switch closes in 0 ms and opens in $6\pi$ ms in the circuit shown. ($t$ is measured in seconds.)

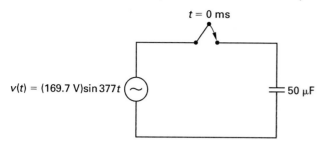

What is most nearly the steady-state value of the total charge on the capacitor?

(A)   0.0 C

(B)   $1.2 \times 10^{-3}$ C

(C)   $6.2 \times 10^{-3}$ C

(D)   170 C

**75.** A voltage of $v(t) = (339.5 \text{ V})\sin 2512t$ is applied to a 30 $\mu$F capacitor that was initially uncharged. ($t$ is measured in seconds.) What is the approximate current in the capacitor at 0.625 ms?

(A)   0.0 A

(B)   0.010 A

(C)   0.020 A

(D)   25 A

**76.** Consider two static charges positioned as shown.

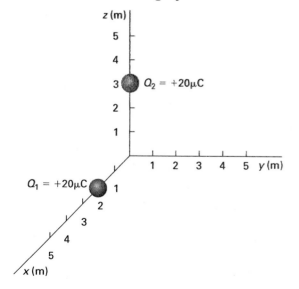

What is most nearly the force on charge $Q_2$?

(A)   $(-0.2 \text{ N})\mathbf{i} + (0.2 \text{ N})\mathbf{k}$

(B)   $(0.5 \text{ N})\mathbf{i} - (0.8 \text{ N})\mathbf{k}$

(C)   $(0.6 \text{ N})\mathbf{i} + (0.8 \text{ N})\mathbf{k}$

(D)   $(-1.8 \text{ N})\mathbf{i} + (2.7 \text{ N})\mathbf{k}$

**77.** What is the approximate magnitude of the electric field 10 m from a 1 C positive point charge in free space?

(A)   1 mV/m

(B)   9 MV/m

(C)   90 MV/m

(D)   900 MV/m

**78.** A uniform current flows through a wire in the direction shown.

Which of the following represents the direction of the magnetic field?

(A)

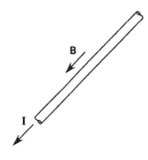

(B)

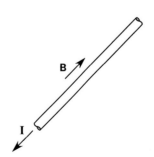

(C)

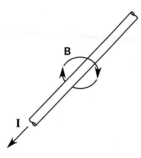

(D)

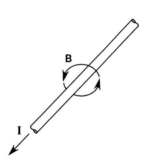

**79.** A digital circuit used to control the lights in a manufacturing center uses three inputs, $A$, $B$, and $C$, to determine the need for lighting. A Karnaugh map of the resulting logic analysis is shown.

| $C$ \ $AB$ | 00 | 01 | 11 | 10 |
|---|---|---|---|---|
| 0 | 1 | 0 | $d$ | 1 |
| 1 | 1 | $d$ | 1 | $d$ |

A programmable logic array (PLA) will be used to realize the function. The unprogrammed PLA is shown.

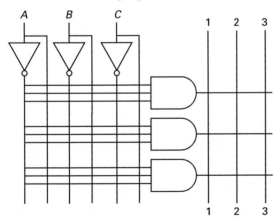

Which of the following PLAs realizes the minimum function, $F(A, B, C)$, on line 1?

(A)

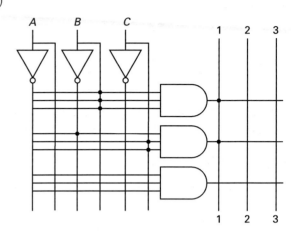

(B)

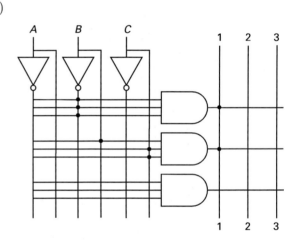

(C)

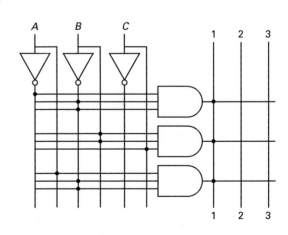

(D)

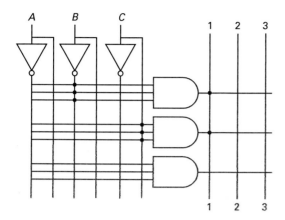

**80.** A manufacturing device is controlled by digital circuitry that uses a three-variable input. The truth table for the input and the required output, $F(X, Y, Z)$, is shown.

| $X$ | $Y$ | $Z$ | $F(X, Y, Z)$ |
|---|---|---|---|
| 0 | 0 | 0 | 0 |
| 0 | 0 | 1 | 1 |
| 0 | 1 | 0 | 0 |
| 0 | 1 | 1 | 1 |
| 1 | 0 | 0 | 0 |
| 1 | 0 | 1 | 1 |
| 1 | 1 | 0 | 0 |
| 1 | 1 | 1 | 1 |

What is the canonical SOP form of the output function?

(A) $\overline{X}\,\overline{Y}Z + \overline{X}YZ + X\overline{Y}Z + XYZ$

(B) $XY\overline{Z} + X\overline{Y}\,\overline{Z} + \overline{X}Y\overline{Z} + \overline{X}\,\overline{Y}\,\overline{Z}$

(C) $(\overline{X} + \overline{Y} + Z)(\overline{X} + Y + Z)(X + \overline{Y} + Z)$
$\times (X + Y + Z)$

(D) $(X + Y + \overline{Z})(X + \overline{Y} + \overline{Z})(\overline{X} + Y + \overline{Z})$
$\times (\overline{X} + \overline{Y} + \overline{Z})$

# STOP!

## DO NOT CONTINUE!

This concludes the Afternoon Session of the examination. If you finish early, check your work and make sure that you have followed all instructions. After checking your answers, you may turn in your examination booklet and answer sheet and leave the examination room. Once you leave, you will not be permitted to return to work or change your answers.

# Answer Keys

## Morning Session Answer Key

| | | | | |
|---|---|---|---|---|
| 1. B | 11. D | 21. A | 31. B |
| 2. B | 12. B | 22. C | 32. D |
| 3. D | 13. D | 23. A | 33. D |
| 4. D | 14. C | 24. B | 34. B |
| 5. A | 15. A | 25. A | 35. D |
| 6. C | 16. A | 26. B | 36. A |
| 7. D | 17. D | 27. C | 37. B |
| 8. C | 18. D | 28. A | 38. A |
| 9. A | 19. A | 29. D | 39. D |
| 10. B | 20. C | 30. B | 40. C |

## Afternoon Session Answer Key

| | | | | |
|---|---|---|---|---|
| 41. D | 51. A | 61. A | 71. A |
| 42. A | 52. C | 62. A | 72. C |
| 43. D | 53. D | 63. C | 73. B |
| 44. A | 54. C | 64. B | 74. C |
| 45. B | 55. D | 65. C | 75. C |
| 46. D | 56. A | 66. B | 76. A |
| 47. C | 57. B | 67. B | 77. C |
| 48. C | 58. A | 68. B | 78. D |
| 49. D | 59. C | 69. B | 79. D |
| 50. A | 60. D | 70. D | 80. A |

# Solutions
## Morning Session Exam

.....................................................................................................................................................................................................

**1.** A vector with the same direction as the magnetic force vector and a magnitude of one is a unit vector. Any unit vector, **a**, is calculated by dividing the vector by its magnitude.

$$\mathbf{a} = \frac{3\mathbf{i} + 6\mathbf{j} + 9\mathbf{k}}{\sqrt{(3)^2 + (6)^2 + (9)^2}}$$

$$= \frac{3\mathbf{i} + 6\mathbf{j} + 9\mathbf{k}}{\sqrt{126}}$$

$$= 0.27\mathbf{i} + 0.53\mathbf{j} + 0.80\mathbf{k}$$

***The answer is (B).***

**2.** The critical points are determined by setting the value of the first derivative to zero and evaluating.

$$y(x) = \sin x - e^x$$
$$y'(x) = \cos x - e^x = 0$$
$$e^x = \cos x$$

The value of $x$ can be determined using numerical techniques or by noting the obvious value for $x$ (i.e., zero), which makes the equation true.

$$e^0 = 1 = \cos 0$$

The value of $y$ at $x = 0$ is

$$y(0) = \sin x - e^x$$
$$= \sin 0 - e^0$$
$$= 0 - 1$$
$$= -1$$

Therefore, a critical point exists in the range $[0, -1]$.

To determine whether the critical point is a local maximum or minimum, find the value of the second derivative at the critical point.

$$y''(x) = -\sin x - e^x$$
$$y''(0) = -\sin 0 - e^0$$
$$= 0 - 1$$
$$= -1$$

When the value of the second derivative at the critical point is greater than zero, the point is a local minimum;

when it is less than zero, the point is a local maximum. In this case, $-1 < 0$, and the point is a local maximum.

The graph of the function is

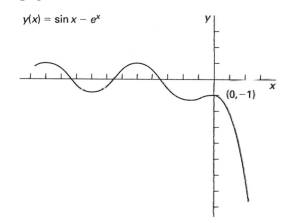

Only one critical point exists in the indicated range, corresponding to the solution.

***The answer is (B).***

**3.** The gradient is

$$\nabla f(x, y, z) = \frac{\partial f(x, y, z)}{\partial x}\mathbf{i} + \frac{\partial f(x, y, z)}{\partial y}\mathbf{j} + \frac{\partial f(x, y, z)}{\partial z}\mathbf{k}$$

$$= \frac{\partial(4x^2 - xy + 3z)}{\partial x}\mathbf{i} + \frac{\partial(4x^2 - xy + 3z)}{\partial y}\mathbf{j}$$

$$+ \frac{\partial(4x^2 - xy + 3z)}{\partial z}\mathbf{k}$$

$$= (8x - y)\mathbf{i} + (-x)\mathbf{j} + 3\mathbf{k}$$

$$= (8x - y)\mathbf{i} - x\mathbf{j} + 3\mathbf{k}$$

***The answer is (D).***

**4.** The transform can be factored into two recognizable inverse Laplace transforms.

$$F(s) = \frac{2}{s^3(s-3)} = \left(\frac{2}{s^3}\right)\left(\frac{1}{s-3}\right) = F_1(s)F_2(s)$$

The inverse transform of $F_1(s)$ is $t^2$. The inverse transform of $F_2(s)$ is $e^{3t}$. Laplace transforms do not commute with ordinary multiplication. That is, $f_1(t)f_2(t) \neq f(t)$, even though $F_1(s)F_2(s) = F(s)$. Instead, convolution is used to determine $f(t)$. (This could also be determined using partial fractions.)

$$f(t) = \int_0^t f_1(\chi)f_2(t-\chi)\,d\chi$$

$$= \int_0^t \chi^2 e^{3(t-\chi)}\,d\chi$$

$$= e^{3t}\int_0^t \chi^2 e^{-3\chi}\,d\chi$$

$$= e^{3t}\left(\left.\frac{\chi^2 e^{-3\chi}}{-3}\right|_0^t - \frac{2}{-3}\int_0^t \chi e^{-3\chi}\,d\chi\right)$$

$$= e^{3t}\left(\frac{t^2 e^{-3t}}{-3} + \left(\frac{2}{3}\right)\left(\left.\frac{e^{-3\chi}}{(-3)^2}(-3\chi-1)\right|_0^t\right)\right)$$

$$= e^{3t}\left(\frac{t^2 e^{-3t}}{-3} + \left(\frac{2}{3}\right)\left(\begin{array}{c}\dfrac{e^{-3t}}{9}(-3t-1) \\[2mm] -\dfrac{e^{(-3)(0)}}{9}\big((-3)(0)-1\big)\end{array}\right)\right)$$

$$= e^{3t}\left(\frac{t^2 e^{-3t}}{-3} + \left(\frac{2}{3}\right)\left(\frac{e^{-3t}}{9}(-3t-1)+\frac{1}{9}\right)\right)$$

$$= e^{3t}\left(\frac{t^2 e^{-3t}}{-3} + \left(\frac{2e^{-3t}}{27}\right)(-3t-1)+\frac{2}{27}\right)$$

$$= e^{3t}\left(\frac{t^2 e^{-3t}}{-3} - \frac{6te^{-3t}}{27} - \frac{2e^{-3t}}{27}+\frac{2}{27}\right)$$

$$= -\frac{t^2}{3} - \frac{2}{9}t - \frac{2}{27} + \frac{2}{27}e^{3t}$$

$$= \frac{2}{27}e^{3t} - \frac{1}{3}t^2 - \frac{2}{9}t - \frac{2}{27}$$

*The answer is (D).*

**5.** Traditionally, problems such as this are worked in the time domain. The use of the phasor concept often simplifies this type of problem and will be used here for edification.

The "ideal conditions" apply to the op amp and also to the discrete elements; that is, the op amp is ideal and the feedback capacitor exhibits no leakage. (Such leakage would make the circuit a "leaky integrator," and the resulting solution, though similar, involves a phase shift.)

Consider node A to be at the negative input to the op amp.

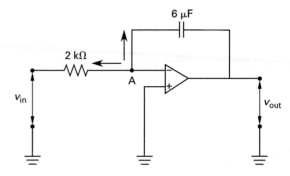

Apply Kirchhoff's current law at node A.

$$\frac{0\text{ V} - \mathbf{V}_{\text{in}}}{\mathbf{Z}_R} + \frac{0\text{ V} - \mathbf{V}_{\text{out}}}{\mathbf{Z}_C} = 0\text{ A}$$

Solve for the output voltage.

$$\mathbf{V}_{\text{out}} = \left(-\frac{\mathbf{Z}_C}{\mathbf{Z}_R}\right)\mathbf{V}_{\text{in}}$$

The term $-\mathbf{Z}_C/\mathbf{Z}_R$ is the gain factor, which includes a negative sign to indicate the inversion of the signal. Substitute the given information, in phasor form, to determine an expression for the output voltage.

$$\mathbf{V}_{\text{out}} = \left(-\frac{\mathbf{Z}_C}{\mathbf{Z}_R}\right)\mathbf{V}_{\text{in}} = \left(-\frac{\dfrac{1}{j\omega C}}{R}\right)\mathbf{V}_{\text{in}}$$

$$= \left(-\frac{1}{j\omega CR}\right)\mathbf{V}_{\text{in}}$$

$$= \left(-\dfrac{1}{\begin{array}{c}(1\angle 90°)(377\text{ Hz}) \\ \times(6\times 10^{-6}\text{ F})(2\times 10^3\ \Omega)\end{array}}\right)(1\text{ V}\angle 0°)$$

$$= 0.221\text{ V}\angle 90°$$

Change this result into the time domain.

$$\mathbf{V}_{\text{out}} = 0.221\text{ V}\angle 90°$$

$$v_{\text{out}} = (0.221\text{ V})\sin(377t+90°) + K$$

$$= (0.221\text{ V})\cos 377t + K$$

$K$ is an arbitrary constant.

Apply the boundary condition $v_{out} = 0$ V at $t = 0$ s to determine the value of $K$.

$$v_{out}(0) = (0.221 \text{ V})\cos 377t + K$$
$$= 0 \text{ V}$$
$$K = (-0.221 \text{ V})\cos\big((377)(0 \text{ V})\big)$$
$$= -0.221 \text{ V}$$

Substitute the value of the constant in the expression for the output voltage. The final solution is

$$v_{out} = (0.221 \text{ V})\cos 377t - 0.221 \text{ V}$$
$$= (0.221 \text{ V})(\cos 377t - 1)$$

**The answer is (A).**

**6.** Using the node voltage method, the circuit currents are as shown.

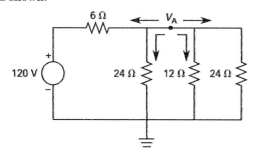

Apply Kirchhoff's current law at node A.

$$\frac{V_A - 120 \text{ V}}{6 \ \Omega} + \frac{V_A}{24 \ \Omega} + \frac{V_A}{12 \ \Omega} + \frac{V_A}{24 \ \Omega} = 0 \text{ A}$$
$$\frac{V_A}{6 \ \Omega} - \frac{120 \text{ V}}{6 \ \Omega} + \frac{V_A}{24 \ \Omega} + \frac{V_A}{12 \ \Omega} + \frac{V_A}{24 \ \Omega} = 0 \text{ A}$$
$$V_A\left(\frac{1}{6 \ \Omega} + \frac{1}{24 \ \Omega} + \frac{1}{12 \ \Omega} + \frac{1}{24 \ \Omega}\right) = \frac{120 \text{ V}}{6 \ \Omega}$$
$$V_A = 60 \text{ V}$$

**The answer is (C).**

**7.** The possible op amp circuits, with their names, are repeated for convenience.

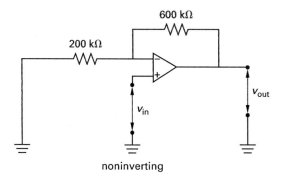

noninverting

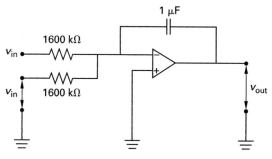

integrator summer

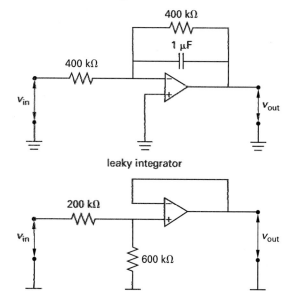

leaky integrator

voltage follower

Assuming ideal op amp properties, only one of the possibilities given has an output resistance of 0 Ω—the circuit in option D. Ensure this is the correct selection by checking the gain of this voltage follower, using the concept of a voltage divider.

$$v_{out} = v^+ = v_{in}\left(\frac{600 \text{ k}\Omega}{800 \text{ k}\Omega}\right)$$
$$= 0.75 v_{in}$$

The gain is 0.75.

The input resistance is the series combination of the 200 kΩ and 600 kΩ resistor, or 800 kΩ as required.

**The answer is (D).**

**8.** The internal resistance of the inductor affects the steady-state time but not the resonant frequency. Accounting for the resistance gives a series RLC circuit as shown.

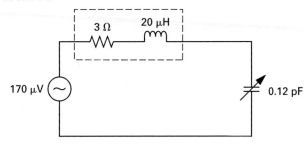

The resonant frequency of a series RLC circuit is

$$f_0 = \frac{1}{2\pi\sqrt{LC}}$$
$$= \frac{1}{2\pi\sqrt{(20 \times 10^{-6} \text{ H})(0.12 \times 10^{-12} \text{ F})}}$$
$$= 102.7 \times 10^6 \text{ Hz} \quad (100 \text{ MHz})$$

**The answer is (C).**

**9.** The wavelength of a given electromagnetic wave in free space is

$$\lambda = \frac{c}{f}$$

$c$ represents the speed of light. The velocity factor, $k$, is the ratio of the speed of the electromagnetic wave in a given transmission line or cable (for example v) to the speed in free space.

$$k = \frac{\text{v}}{c}$$

Combining the two equations gives the wavelength of an electromagnetic wave in a cable with velocity factor $k$.

$$\lambda = \frac{kc}{f}$$

Substitute the given information.

$$\lambda = \frac{(0.6)\left(3 \times 10^8 \ \frac{\text{m}}{\text{s}}\right)}{1 \times 10^9 \text{ Hz}}$$
$$= 0.18 \text{ m} \quad (0.2 \text{ m})$$

**The answer is (A).**

**10.** The period of a given waveform is

$$T = \frac{1}{f}$$

The frequency is

$$\omega = 2\pi f$$
$$f = \frac{\omega}{2\pi} = \frac{2513 \text{ Hz}}{2\pi}$$
$$= 400 \text{ Hz}$$

Substitute the calculated value to determine the period.

$$T = \frac{1}{f} = \frac{1}{400 \text{ Hz}}$$
$$= 2.5 \times 10^{-3} \text{ s} \quad (2.5 \text{ ms})$$

**The answer is (B).**

**11.** This solution involves the complex frequency $\mathbf{s} = \sigma + j\omega$. Use only the real portion of $e^{jxt} = \cos xt + j \sin xt$, given a constant $A$ and factor $e^{\sigma t}$.

$$A e^{\sigma t} e^{j(\omega t + \phi)} = A e^{\sigma t} e^{j\omega t} e^{j\phi}$$
$$= A e^{j\omega t + \sigma t} e^{j\phi}$$
$$A e^{(\sigma + j\omega)t} e^{j\phi} = A e^{st} e^{j\phi}$$
$$= A e^{j\phi} e^{st}$$
$$\text{Re}\left(A e^{\sigma t} e^{j(\omega t + \phi)}\right) = A e^{\sigma t} \cos(\omega t + \phi)$$

Using the relationships in the listed equations, consider $3e^{-2t} \cos(377t - 30°)$. The constant term $A$ is 3 and the term $\phi$ is 30°, neither of which is included in $\mathbf{s}$, since these terms represent values that do not change with time. The term $\sigma$, which is also called the neper frequency, is $-2$ Hz. The angular frequency $\omega$ is 377 Hz. The complex frequency, which is used to represent the time-varying portions of the signal, is

$$\mathbf{s} = -2 + j377$$

The complex frequency $\mathbf{s}$ is usually written as $s$ when used in the exponent, $e^{st}$.

**The answer is (D).**

**12.** Unless specified otherwise, op amps may be treated as ideal. Consider the circuit with the currents at the inverting terminal arbitrarily defined as shown.

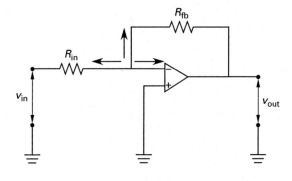

Apply Kirchhoff's current law at the inverting terminal.

$$\frac{0\text{ V} - v_{in}}{R_{in}} + \frac{0\text{ V} - v_{out}}{R_{fb}} + 0\text{ A} = 0\text{ A}$$

$$\frac{v_{out}}{R_{fb}} = \frac{-v_{in}}{R_{in}}$$

$$v_{out} = v_{in}\left(-\frac{R_{fb}}{R_{in}}\right)$$

The output is inverted. The gain is set by the ratio of the feedback resistor value and the input resistor value.

*The answer is (B).*

**13.** The op amps are used to compare the input voltage on the positive (+) terminal to the reference voltage on the negative (−) terminal. The output of the op amp drives to the positive supply voltage (+5 V) if the positive input terminal voltage magnitude is higher, and to the negative supply voltage (−5 V) if the negative input terminal voltage magnitude is higher.

The negative input terminal voltage is set by the resistor network as shown.

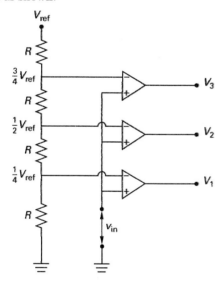

The voltage at $\frac{3}{4}V_{ref}$ is 0.75 V. The input voltage is 0.76 V. Therefore, the voltage on the positive input terminal is greater than the voltage on the negative input terminal on all three op amps. The output of all the op amps will be at the positive power-supply voltage level of +5 V, or HIGH (H, H, H).

*The answer is (D).*

**14.** Assume the op amp can be accurately approximated as ideal, which is normally the case. The directions used to apply Kirchhoff's current law (KCL) at the negative input terminal are

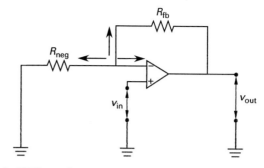

Apply KCL at the negative input terminal.

$$\frac{v_{in} - 0\text{ V}}{R_{neg}} + \frac{v_{in} - v_{out}}{R_{fb}} + 0\text{ A} - 0\text{ A}$$

Rearrange the equation to solve for the output voltage expression.

$$\frac{v_{out}}{R_{fb}} = \frac{v_{in}}{R_{neg}} + \frac{v_{in}}{R_{fb}}$$

$$v_{out} = v_{in}\left(\frac{R_{fb}}{R_{neg}} + 1\right)$$

$$= v_{in}\left(1 + \frac{R_{fb}}{R_{neg}}\right)$$

The output is not inverted, and the gain is greater than 1 for this configuration.

*The answer is (C).*

**15.** Assume the op amp is ideal.

$$v^- = v^+$$

The voltages at the input terminals are

$$v^- = v_{out}$$
$$v^+ = v_{in}$$

Given the observation that the input terminal voltages are equal,

$$v_{out} = v_{in}$$

The loading effect of the load resistor on the source is eliminated, hence the name "buffer" for the op amp configuration. (This is also called a voltage follower, since the output voltage follows the input voltage.)

*The answer is (A).*

**16.** The voltage across the inductor is

$$v(t) = L\frac{di}{dt} = L\frac{d(8 \text{ A})\sin 25t}{dt}$$

$$= L(8 \text{ A})\cos 25t(25 \text{ Hz})$$

$$= L\left(200 \ \frac{\text{A}}{\text{s}}\right)\cos 25t$$

Substitute the given information.

$$L\left(200 \ \frac{\text{A}}{\text{s}}\right)\cos 25t = (90 \times 10^{-3} \text{ H})\left(200 \ \frac{\text{A}}{\text{s}}\right)$$

$$\times \cos\left((25)\left(\frac{\pi}{50}\right)\right)$$

$$= (18 \text{ V})\cos\frac{\pi}{2}$$

$$= (18 \text{ V})(0)$$

$$= 0 \text{ V} \quad (0.0 \text{ V})$$

*The answer is (A).*

**17.** Use the mesh current method with loop currents as shown.

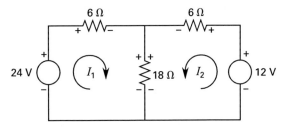

Apply Kirchhoff's voltage law (KVL) in loop 1.

$$24 \text{ V} - I_1(6 \text{ }\Omega) - (I_1 + I_2)(18 \text{ }\Omega) = 0 \text{ V}$$

Rearrange.

$$I_1 + \frac{3}{4}I_2 = 1 \text{ A} \qquad \text{[I]}$$

Apply KVL in loop 2.

$$12 \text{ V} - I_2(6 \text{ }\Omega) - (I_1 + I_2)(18 \text{ }\Omega) = 0 \text{ V}$$

Rearrange.

$$I_1 + \frac{4}{3}I_2 = 2/3 \text{ A} \qquad \text{[II]}$$

Solve Eqs. I and II simultaneously.

$$I_1 = 10/7 \text{ A}$$

$$I_2 = -4/7 \text{ A}$$

The total current in the 18 $\Omega$ resistor is the sum of the loop currents.

$$I_t = I_1 + I_2 = \frac{10}{7} \text{ A} + \left(-\frac{4}{7} \text{ A}\right)$$

$$= 6/7 \text{ A}$$

*The answer is (D).*

**18.** A circuit is considered to be in steady state once five time constants have elapsed. The time constant for a single-resistor, single-capacitor circuit is

$$\tau = RC = (600 \text{ }\Omega)(75 \times 10^{-6} \text{ F})$$

$$= 0.045 \text{ s}$$

Determine the passage of five time-constant periods.

$$5\tau = (5)(0.045 \text{ s})$$

$$= 0.225 \text{ s} \quad (0.22 \text{ s})$$

*The answer is (D).*

**19.** The current cannot be calculated using resistance values, since none are given. Consider the circuit redrawn as shown.

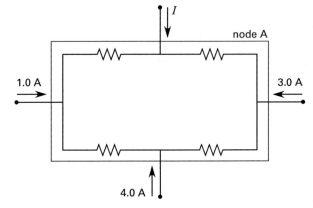

The resistor network is treated as a single node, A. (The −4.0 A illustrated in the problem statement indicates that the flow of current is into the node.) Apply Kirchhoff's current law.

$$\sum_{\text{node}} I_{\text{in}} = \sum_{\text{node}} I_{\text{out}}$$

$$1.0 \text{ A} + 3.0 \text{ A} + I + 4.0 \text{ A} = 0 \text{ A}$$

Rearrange to solve for $I$.

$$I = 0 \text{ A} - 1.0 \text{ A} - 3.0 \text{ A} - 4.0 \text{ A}$$

$$= -8.0 \text{ A}$$

*The answer is (A).*

**20.** The charge on the capacitor is found using the following fundamental relationship.

$$Q = CV$$
$$= (15 \times 10^{-6} \text{ F})(90 \text{ V})$$
$$= 1.35 \times 10^{-3} \text{ C} \quad (1.4 \text{ mC})$$

*The answer is (C).*

**21.** The Norton equivalent circuit is defined by the open-circuit voltage and the short-circuit current as shown. (The resistors have been identified with subscripts for clarification.)

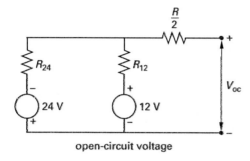

open-circuit voltage

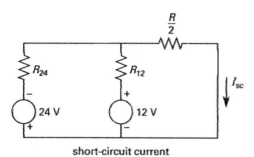

short-circuit current

The solution can be determined using a number of different methods. Using superposition, the current from the 12 V source is

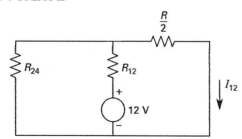

Using the concept of a current divider, $I_{12}$ is

$$I_{12} = \left(\frac{12 \text{ V}}{R_{12} + \dfrac{R_{24}\left(\dfrac{R}{2}\right)}{R_{24} + \dfrac{R}{2}}}\right)\left(\frac{R_{24}}{R_{24} + \dfrac{R}{2}}\right)$$

$$= \left(\frac{12 \text{ V}}{R + \dfrac{R}{3}}\right)\left(\frac{R}{\frac{3}{2}R}\right)$$

$$= \frac{6 \text{ V}}{R}$$

Using superposition, the current from the 24 V source is

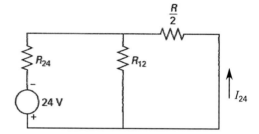

Using the concept of a current divider, $I_{24}$ is

$$I_{24} = \left(\frac{24 \text{ V}}{R_{24} + \dfrac{R_{12}\left(\dfrac{R}{2}\right)}{R_{12} + \dfrac{R}{2}}}\right)\left(\frac{R_{12}}{R_{12} + \dfrac{R}{2}}\right)$$

$$= \left(\frac{24 \text{ V}}{R + \dfrac{R}{3}}\right)\left(\frac{R}{\frac{3}{2}R}\right)$$

$$= \frac{12 \text{ V}}{R}$$

The short-circuit current, as defined, is

$$I_{SC} = I_{12} - I_{24} = \frac{6 \text{ V}}{R} - \frac{12 \text{ V}}{R} = -\frac{6 \text{ V}}{R}$$

$$= -\frac{6 \text{ V}}{2 \ \Omega}$$

$$= -3 \text{ A}$$

The resulting circuit is

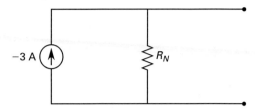

The magnitude of the current is 3 A.

*The answer is (A).*

**22.** The triangular waveform has even symmetry (i.e., $f(-t) = f(t)$), so option A is true. Even symmetry results in a series with cosine terms and no sine terms, meaning the $b_n$ coefficients associated with the sine terms are 0 and options B and D are true.

The first term in a Fourier series is the zero-frequency term, or the average term, and it is zero only if the average value is zero, which is clearly not the case for a function with only positive value.

*The answer is (C).*

**23.** The fundamental properties of an inductor can be modeled as a short circuit for low frequencies (DC) and an open circuit for high frequencies. The fundamental properties of a capacitor can be modeled as an open circuit for low frequencies and a short circuit for high frequencies.

For both the high- and low-frequency cases, one of the two circuit elements in the input circuitry would act as an open circuit, preventing any of the input voltage from reaching the load resistor, $R$. That is, the normalized input impedance is high at both low- and high-frequency levels. The only frequency response that meets these conditions is option A, which is the response of a circuit displaying series resonance.

*The answer is (A).*

**24.** Transforming the individual circuit elements into the $s$-domain and accounting for the initial capacitor voltage results in the circuit shown.

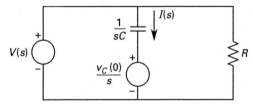

Use Ohm's law across the capacitor, rearranging for the current.

$$
\begin{aligned}
I(s) &= \frac{V(s) - \dfrac{v_C(0)}{s}}{\dfrac{1}{sC}} \\
&= sCV(s) - Cv_C(0) \\
&= C\big(sV(s) - v_C(0)\big)
\end{aligned}
$$

*The answer is (B).*

**25.** An accelerating charge generates a steadily radiating electromagnetic wave.

*The answer is (A).*

**26.** Of all the possible combinations, ii and vi (across a current-free interface) represent the properties that are continuous.

*The answer is (B).*

**27.** The intrinsic impedance of free space is

$$
\begin{aligned}
\eta_0 &= \sqrt{\frac{\mu_0}{\epsilon_0}} \\
&= \sqrt{\frac{1.257 \times 10^{-6} \ \dfrac{\text{H}}{\text{m}}}{8.854 \times 10^{-12} \ \dfrac{\text{F}}{\text{m}}}} \\
&= 376.8 \ \Omega \quad (377 \ \Omega)
\end{aligned}
$$

The permeability and permittivity of free space are sometimes represented in terms of $\pi$, as shown. The result is the same.

$$
\begin{aligned}
\eta_0 = \sqrt{\frac{\mu_0}{\epsilon_0}} &\approx \sqrt{\frac{4\pi \times 10^{-7} \ \dfrac{\text{H}}{\text{m}}}{\dfrac{10^{-9}}{36\pi} \ \dfrac{\text{F}}{\text{m}}}} \\
&= 120\pi \ \Omega \\
&= 377 \ \Omega
\end{aligned}
$$

*The answer is (C).*

**28.** Since the field direction does not vary, the magnitude of the displacement current density is

$$
\begin{aligned}
J_D &= \frac{\partial D}{\partial t} = \frac{\partial(\epsilon_0 \epsilon_r E)}{\partial t} \\
&= \left(175\ \frac{\mathrm{V}}{\mathrm{m}}\right) \epsilon_0 \epsilon_r \frac{\partial \sin(10)^9 t}{\partial t} \\
&= \left(175\ \frac{\mathrm{V}}{\mathrm{m}}\right)\left(8.854 \times 10^{-12}\ \frac{\mathrm{F}}{\mathrm{m}}\right) \\
&\quad \times (1)\big((10)^9\ \mathrm{Hz}\big)\cos 10^9 t \\
&= \left(1.55\ \frac{\mathrm{A}}{\mathrm{m}^2}\right)\cos 10^9 t
\end{aligned}
$$

**The answer is (A).**

**29.** Option A represents free-space conditions in the mathematical point-form of Maxwell's equations.

Option B is the general statement in mathematical point-form of Gauss' law. That is, the total flux out of a closed surface is equal to the net charge enclosed.

Option C represents the nonexistence of magnetic monopoles (in both the general and free-space sets of Maxwell's equations).

Therefore, option D is the only false statement.

**The answer is (D).**

**30.** The overall gain, $K$, without feedback is

$$
K = K_1 K_2 K_3 = (-50)(-75)(-80) = -300\,000
$$

The fraction of the output signal appearing at the summing point, $h$, depends on the resistances of the voltage divider.

$$
h = \frac{R_{25}}{R_{25} + R_{1000}} = \frac{25\ \Omega}{25\ \Omega + 1000\ \Omega} = 0.024
$$

The feedback is summed (i.e., adds to the input), so the positive feedback formula for gain is used. But since the gain is itself negative ($K < 0$), the overall feedback is negative.

$$
\begin{aligned}
K_{\mathrm{loop}} &= \frac{K}{1 - Kh} = \frac{-300\,000}{1 - (-300\,000)(0.024)} \\
&= -41.66 \quad (-40)
\end{aligned}
$$

**The answer is (B).**

**31.** The denominator of the transfer function exhibits the general form of the characteristic equation of a higher-order circuit.

$$
as^2 + bs + c
$$

For control systems, the characteristic equation is

$$
s^2 + 2\zeta \omega_n s + \omega_n^2
$$

For resonance conditions, this equation takes the form

$$
s^2 + \left(\frac{\omega_0}{Q}\right)s + \omega_0^2
$$

Compare the denominator of the given transfer function with the resonant form of the characteristic equation. Note that the magnitude of the resonant frequency squared is $10^6$, making the resonant frequency magnitude $10^3$.

$$
20\ \frac{\mathrm{rad}}{\mathrm{s}} = \frac{\omega_0}{Q}
$$

$$
Q = \frac{\omega_0}{20\ \dfrac{\mathrm{rad}}{\mathrm{s}}} = \frac{10^3\ \dfrac{\mathrm{rad}}{\mathrm{s}}}{20\ \dfrac{\mathrm{rad}}{\mathrm{s}}}
$$

$$
= 50
$$

**The answer is (B).**

**32.** Using block diagram algebra, the feedback loops can be simplified to the following.

The gain, $C$, can be combined with either of the other two blocks. Choose the leftmost block.

The remaining two blocks are combined to give the transfer function (i.e., the gain) of the circuit.

**The answer is (D).**

**33.** The forcing function is a step of height 2 at $t = 0$ s. The Laplace transform of the unit step is $1/s$. Therefore, $F(s) = 2/s$. The response function is

$$
\begin{aligned}
R(s) &= F(s)\,T(s) = \left(\frac{2}{s}\right)\left(\frac{6}{s+2}\right) \\
&= \frac{12}{s(s+2)}
\end{aligned}
$$

Using partial fractions, the response function can be represented as

$$R(s) = \frac{12}{s(s+2)}$$
$$= \frac{6}{s} - \frac{6}{s+2}$$

The time-based response function is found using the inverse Laplace transform of the response function.

$$r(t) = \mathcal{L}^{-1}\left(\frac{6}{s} - \frac{6}{s+2}\right)$$
$$= 6 - 6e^{-2t}$$

*The answer is (D).*

**34.** The final value of the time-based $r(t)$ is found from the frequency based $R(s)$ using the final value theorem, which states

$$\lim_{t\to\infty} r(t) = \lim_{s\to 0}\left(sR(s)\right)$$

Multiply $R(s)$ by $s$ and apply the limit.

$$r(\infty) = \lim_{s\to 0}\left(sR(s)\right) = \lim_{s\to 0}\frac{s}{s+1}$$
$$= \lim_{s\to 0}\frac{1}{s+1}$$
$$= 1$$

*The answer is (B).*

**35.** All the statements are true with the exception of option D. The bucket brigade is a method for approximating the dead time in a given controller.

*The answer is (D).*

**36.** The numerator is zero when $s=-4$. This is the only zero in the transfer function. (Thus, option A should be the correct answer.)

The denominator is zero when $s=-2$ and $s=-1\pm j$. The later zero can be seen by factoring the quadratic in the denominator. These three values are the poles of the transfer function.

*The answer is (A).*

**37.** The phase margin is the number of degrees the phase angle is above $-180°$ at the gain crossover point (i.e., where the logarithmic gain is 0 dB or the actual gain is 1). This is point B.

*The answer is (B).*

**38.** The reflection coefficient of the transmission line is

$$\Gamma_L = \frac{Z_L - Z_0}{Z_L + Z_0} = \frac{120\ \Omega - 60\ \Omega}{120\ \Omega + 60\ \Omega}$$
$$= 0.33$$

*The answer is (A).*

**39.** The effective isotropically radiated power (EIRP) is

$$\text{EIRP} = 10\log G_T P_T$$

The subscript $T$ indicates the transmitting antenna. The gain for the transmitting antenna is given in dBW, and must be converted to a numerical value as shown.

$$G_{T,\text{dBW}} = 10\log G_T$$
$$17 = 10\log G_T$$
$$1.7 = \log G_T$$
$$G_T = \text{antilog } 1.7 \approx 50$$

Substitute this value for the gain into the equation for the EIRP.

$$\text{EIRP} = 10\log(G_T P_T) = 10\log\frac{(50)(2\text{ W})}{1\text{ W}}$$
$$= 20\text{ dBW}$$

*The answer is (D).*

**40.** The modulated FM signal, in terms of its constituent parts, is

$$s_{\text{mod}}(t) = A\cos\left(\omega_c t + \left(\frac{\Delta\omega}{\omega_{\text{mod}}}\right)\sin\omega_{\text{mod}}t\right)$$

Start with the index of modulation, $m_{\text{FM}}$, which is the term $\Delta\omega/\omega_{\text{mod}}$, and substitute the given information to solve for $\Delta f$.

$$m_{\text{FM}} = \frac{\Delta\omega}{\omega_{\text{mod}}}$$
$$8 = \frac{2\pi\Delta f}{10^6\ \dfrac{\text{rad}}{\text{s}}}$$
$$\Delta f = \frac{8\times 10^6\ \dfrac{\text{rad}}{\text{s}}}{2\pi}$$
$$= 1.27\times 10^6\text{ Hz}\quad(1.3\times 10^6\text{ Hz})$$

In terms of the coherent SI system, the unit "radian" has a dimension of 1. The term "rad" is used to aid understanding only.

*The answer is (C).*

# Solutions
## Afternoon Session Exam

**41.** The figure in option A is the piecewise linear model for a diode. The figure in option B is an *h*-parameter, small-signal, simplified equivalent circuit for a common-emitter bipolar junction transistor (BJT). The figure in option C is an *h*-parameter, small-signal, simplified equivalent circuit for a common-collector BJT. The figure in option D is a simplified model of a field-effect transistor. (The simplifications in each model consist of ignoring parameters that are insignificant; for example, extremely small reverse currents or near-infinite parallel resistance.)

**The answer is (D).**

**42.** The work done on the particle, $W$, is the dot product of the force, $\mathbf{F}$, and the distance, $\mathbf{s}$.

$$
\begin{aligned}
W &= \mathbf{F} \cdot \mathbf{s} \\
&= (2\mathbf{i} + 3\mathbf{j} + \mathbf{k}) \cdot \big((-1 - 0)\mathbf{i} + (3 - 3)\mathbf{j} + (-1 - 1)\mathbf{k}\big) \\
&= (2\mathbf{i} + 3\mathbf{j} + \mathbf{k}) \cdot (-1\mathbf{i} + 0\mathbf{j} - 2\mathbf{k}) \\
&= (2)(-1)(\mathbf{i} \cdot \mathbf{i}) + (3)(0)(\mathbf{j} \cdot \mathbf{j}) + (1)(-2)(\mathbf{k} \cdot \mathbf{k}) \\
&= -2 + 0 - 2 \\
&= -4
\end{aligned}
$$

The negative work means the force and displacement are in opposite directions; that is, work is done on the item providing the force. Mathematically, this means the angle between the vectors is greater than $90°$, or approximately $120°$ in this case.

**The answer is (A).**

**43.** The following figure shows a graphical representation of complex numbers and the equation for the imaginary component $j$.

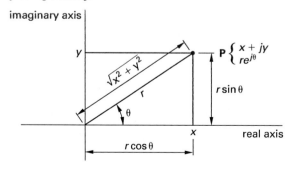

Rectangular forms of complex numbers separate the real and imaginary components into their axis values. Polar forms use the radius and angle components. If the cosine and sine nomenclature is used, the form is called the trigonometric form. Using $e$ in the representation makes it the exponential form. Finally, the complex number can also be thought of as a vector from the origin (the intersection of the $x$- and $y$-axes) to the point P.

Change the rectangular form into the exponential form using the following conversions.

$$
\begin{aligned}
r &= \sqrt{x^2 + y^2} \\
&= \sqrt{(3)^2 + (4)^2} \\
&= 5 \\
\theta &= \arctan \frac{y}{x} \\
&= \arctan \frac{4}{3} \\
&= 0.927 \text{ rad} \\
\mathrm{P} &= re^{j\theta} \\
&= 5e^{j0.927}
\end{aligned}
$$

Euler's equation is

$$
e^{j\theta} = \cos\theta + j\sin\theta
$$

Use Euler's equation to change the exponential form into one using trigonometric functions.

$$
\begin{aligned}
\mathrm{P} &= 5e^{j0.927} \\
&= (5)(\cos 0.927 + j\sin 0.927) \\
&\quad \big((5)(\cos 0.93 + j\sin 0.93)\big)
\end{aligned}
$$

**The answer is (D).**

**44.** The op amp will respond to the instantaneous values of the voltage. The relationship between the peak voltage and the root-mean-square voltage for a sinusoidal voltage is

$$
\begin{aligned}
|V_{\text{peak}}| &= \sqrt{2}\, V_{\text{rms}} \\
&= \sqrt{2}\,(250 \text{ V}) \\
&= 353.55 \text{ V}
\end{aligned}
$$

The op amp is not able to follow the input voltage more than 13 V (positive or negative). Choose to make the op amp input peak voltage of 13 V (which is 3 V less than the power supply voltage needed to ensure linear operation) correspond to the peak voltage (both positive and negative) of 353.55 V, that is, 250 $V_{rms}$. The voltage divider shown attached to the positive terminal accomplishes the adjustment if the value of $R_{div}$ is

$$(353.55 \text{ V})\left(\frac{R_{div}}{R_{div} + 1 \times 10^6 \ \Omega}\right) = 13 \text{ V}$$

$$(353.55 \text{ V})R_{div} = (13 \text{ V})(R_{div} + 1 \times 10^6 \ \Omega)$$
$$= (13 \text{ V})(R_{div}) + 13 \times 10^6 \text{ V·}\Omega$$

$$(353.55 \text{ V})R_{div} - (13 \text{ V})R_{div} = 13 \times 10^6 \text{ V·}\Omega$$
$$R_{div}(353.55 \text{ V} - 13 \text{ V}) = 13 \times 10^6 \text{ V·}\Omega$$
$$R_{div} = \frac{13 \times 10^6 \text{ V·}\Omega}{340.55 \text{ V}}$$
$$= 38 \times 10^3 \ \Omega \quad (40 \text{ k}\Omega)$$

*The answer is (A).*

**45.** The equation for the current through the inductor is

$$v_L(t) = L\left(\frac{di(t)}{dt}\right)$$

Substituting the given current yields

$$v_L(t) = L\left(\frac{d(1.5 \text{ A})(1 - e^{-1000t})}{dt}\right)$$
$$= L\left(1.5 \ \frac{\text{A}}{\text{s}}\right)(1000e^{-1000t})$$
$$= (5.0 \times 10^{-3} \text{ H})\left(1.5 \ \frac{\text{A}}{\text{s}}\right)(1000e^{-1000t})$$
$$= 7.5e^{-1000t} \text{ V}$$

At 100 $\mu$s into a transient, the voltage is

$$v_L(t) = 7.5e^{-1000t} \text{ V}$$
$$= 7.5e^{-(1000)(100 \times 10^{-6} \text{ s})} \text{ V}$$
$$= 7.5e^{-0.1} \text{ V}$$
$$= 6.8 \text{ V}$$

*The answer is (B).*

**46.** The work done in moving a point charge within an electrostatic field is

$$dW = -Q\mathbf{E} \cdot d\mathbf{l}$$

Given that the path is entirely on the x-axis, $d\mathbf{l} = dx\mathbf{i}$. Substitute.

$$dW = -Q\mathbf{E} \cdot dx\mathbf{i}$$
$$= -(-1.6 \times 10^{19} \text{ C})(2x + 4y)dx$$

Integrate along the x-axis from 0 to 4. The electric field units are N/C, and the integration takes place over the x-axis units of meters.

$$W = \int_0^4 dW = -Q\int_0^4 \mathbf{E} \cdot dx \ \mathbf{i}$$
$$= -(-1.6 \times 10^{19} \text{ C})\int_0^4 (2x + 4y)dx$$
$$= (1.6 \times 10^{19} \text{ C})\left(\int_0^4 2x \ dx + 4y\int_0^4 dx\right)$$
$$= (1.6 \times 10^{19} \text{ C})\left([x^2]\Big|_0^4 + 4(0)\int_0^4 dx\right)$$
$$= (1.6 \times 10^{19} \text{ C})\left(16 \ \frac{\text{N·m}}{\text{C}}\right)$$
$$= 2.56 \times 10^{20} \text{ J}$$

*The answer is (D).*

**47.** The electric flux density, **D**, is

$$\mathbf{D} = \epsilon_0\epsilon_r\mathbf{E}$$

Rearrange to solve for the electric field strength and substitute the given information.

$$\mathbf{E} = \frac{\mathbf{D}}{\epsilon_0\epsilon_r} = \frac{1.5 \times 10^{-6}\mathbf{a}_r \ \frac{\text{C}}{\text{m}^2}}{\left(8.854 \times 10^{-12} \ \frac{\text{C}^2}{\text{N·m}^2}\right)(2.9)}$$
$$= 5.84 \times 10^4\mathbf{a}_r \text{ N/C} \quad (5.8 \times 10^4\mathbf{a}_r \text{ N/C})$$

*The answer is (C).*

**48.** Find the impedance by dividing the voltage by the current.

$$Z = \frac{V}{I} = \frac{3 + j5 \text{ V}}{4 - j10 \text{ A}}$$

For complex numbers, the numerator and denominator are multiplied by the complex conjugate of the denominator.

$$Z = \left(\frac{3 + j5 \text{ V}}{4 - j10 \text{ A}}\right)\left(\frac{4 + j10}{4 + j10}\right)$$

$$= \frac{12 + j^2 50 + j20 + j30}{16 - j^2 100} \; \Omega$$

$$= \frac{12 - 50 + j50}{16 + 100} \; \Omega$$

$$= -0.33 + j0.43 \; \Omega$$

**The answer is (C).**

**49.** Obtain the Thevenin equivalent of the gate-source circuit.

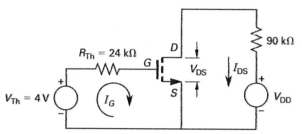

Write the KVL around the gate loop as indicated.

$$V_{\text{Th}} - I_G R_{\text{Th}} - V_{\text{GS}} = 0$$

The gate current in a MOSFET is zero amperes. Thus,

$$V_{\text{GS}} = V_{\text{Th}} = 4.0 \text{ V}$$

**The answer is (D).**

**50.** The critical incident angle, $\theta_c$, for a wave incident upon a surface free space, which air approximates, is

$$\sin\theta_c = \frac{1}{n}$$

Rearrange and substitute the known value for the absolute index of refraction.

$$\theta_c = \arcsin\frac{1}{n}$$

$$= \arcsin\frac{1}{2}$$

$$= 30° \quad (\pi/6 \text{ rad})$$

**The answer is (A).**

**51.** When a receiving antenna is located to receive the maximum power density from a transmitting antenna, the power density available is

$$p_r = \left(\frac{D_T}{4\pi r^2}\right)P_T$$

$$= \left(\frac{1.5}{4\pi(5 \times 10^3 \text{ m})^2}\right)(100 \times 10^3 \text{ W})$$

$$= 4.77 \times 10^{-4} \text{ W/m}^2 \quad (0.5 \text{ mW/m}^2)$$

**The answer is (A).**

**52.** An open-loop transfer function for $v_2$ is one in which only the inputs $(v_1, R_1, R_2)$ determine the output $(v_2)$. From Ohm's law and Kirchhoff's laws,

$$v_{\text{load}} = iR_2$$

$$i = \frac{v_{\text{source}}}{R_1 + R_2}$$

Combining the two such that the load voltage is determined only by the inputs gives the open-loop transfer function.

$$v_{\text{load}} = \left(\frac{R_2}{R_1 + R_2}\right)v_{\text{source}}$$

Option D is the closed-loop function for $v_{\text{load}}$ because the output is taken into account, that is, it is fed back to the input that determines the actual output.

**The answer is (C).**

**53.** The open-loop transfer function is given by $GH$. The lay reasoning is that the signal has not completed the loop. The closed-loop transfer function is $C/R$ since the feedback transfer function ($H$) has been taken into account in the output.

**The answer is (D).**

**54.** The input signal is on the emitter. The output signal is taken from the collector. Both share the common base.

A common base configuration is generally known for the following properties: a high (positive) voltage gain, $A_V$, corresponding to option C; a low current gain, $A_I$; a low input resistance, $R_i$; and a high output resistance, $R_o$.

**The answer is (C).**

**55.** To know where the transistor saturates, calculate the $Q$-point and use this information to determine the magnitude of the input signal necessary to cause saturation. The load line for this circuit, which is not a necessary part of the solution, is shown to aid understanding.

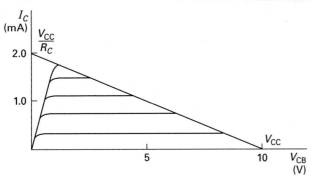

The biasing currents are shown.

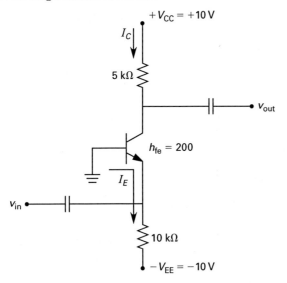

First, calculate the controlling emitter current using Kirchhoff's voltage law (KVL) around the emitter loop.

$$-V_{BE} - V_E + V_{EE} = 0 \text{ V}$$

Rearrange and solve for the emitter current.

$$V_E = V_{EE} - V_{BE}$$
$$I_E R_E = V_{EE} - V_{BE}$$
$$I_E = \frac{V_{EE} - V_{BE}}{R_E}$$

For a silicon transistor, the base-emitter voltage drop is approximately 0.7 V. (The value of $V_{EE}$ is 10 V. The

negative sign was accounted for in the KVL equation.) Substitute the known and given information.

$$I_E = \frac{V_{EE} - V_{BE}}{R_E}$$
$$= \frac{10 \text{ V} - 0.7 \text{ V}}{10 \times 10^3 \text{ } \Omega}$$
$$= 9.3 \times 10^{-4} \text{ A}$$

The collector current corresponding to this emitter current is

$$I_C \approx \alpha I_E = \left(\frac{\beta}{1 + \beta}\right) I_E$$

The $\alpha$ and $\beta$ values are DC quantities. Nevertheless, the value of the current amplification ratio, $\beta$, can be determined from the given value for $h_{fe}$, which is an AC quantity.

$$h_{fe} = \beta_{ac} \approx \beta_{DC}$$
$$\alpha_{DC} = \frac{\beta_{DC}}{1 + \beta_{DC}} \approx \frac{h_{fe}}{1 + h_{fe}}$$
$$= \frac{200}{1 + 200}$$
$$= 9.95 \times 10^{-1}$$

(The distinction between $\beta_{ac}$ and $\beta_{DC}$ is seldom made but is shown here for clarification.) Substitute the value of $\alpha$ and the calculated emitter current.

$$I_C \approx \alpha I_E$$
$$= (9.95 \times 10^{-1})(9.3 \times 10^{-4} \text{ A})$$
$$= 9.25 \times 10^{-4} \text{ A}$$

The output loop for this circuit is shown.

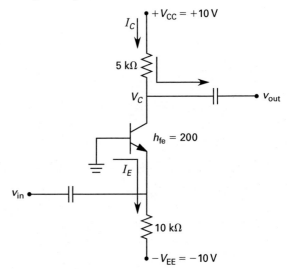

Write KVL in the output loop and rearrange to determine the collector voltage, $V_C$, at the $Q$-point.

$$\begin{aligned} V_C &= V_{CC} - I_C R_C \\ &= 10 \text{ V} - (9.25 \times 10^{-4} \text{ A})(5 \times 10^3 \text{ }\Omega) \\ &= 5.38 \text{ V} \quad (5.4 \text{ V}) \end{aligned}$$

Plot the $Q$-point for clarification.

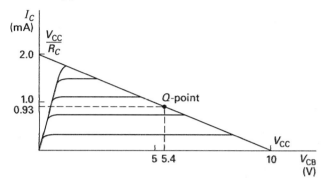

A negative swing of 5.4 V on the input will cause saturation. (A positive swing of 4.6 V will result in the cutoff state.)

*The answer is (D).*

**56.** The collector-base cutoff current—essentially, the reverse saturation current—doubles with every 10°C rise. The thermal stability is affected by this current. The effect is self-reinforcing: as the temperature increases, the saturation current also increases. This is called thermal runaway. The emitter resistor helps stabilize the transistor against this trend. As the current rises, the voltage drop across the emitter resistor rises in a direction that opposes the forward biased base-emitter junction.

*The answer is (A).*

**57.** Control systems use the following form for the characteristic equation.

$$s^2 + 2\zeta\omega_n s + \omega_n^2$$

The damping ratio is represented by the symbol $\zeta$. Solve for this value using the given information.

$$\begin{aligned} 2\zeta\omega_n &= 4 \\ \zeta &= \frac{4}{2\omega_n} \\ &= \frac{4}{2\sqrt{144}} \\ &= 0.17 \end{aligned}$$

*The answer is (B).*

**58.** Commercial AM broadcasts are generally of the double-sideband large carrier (DSB-LC) type—option D is true. Both sidebands are used—option C is true. These sidebands can be determined using Fourier analysis, which shows the signal as frequency shifted from the original carrier—option B is true.

Suppressing a carrier signal requires receivers to maintain synchronization and increases the cost, so the carrier signal remains in the commercial AM broadcasts.

*The answer is (A).*

**59.** All the polynomial terms are present, satisfying the Hurwitz criterion. The Routh table, constructed from the characteristic equation, is

| | | |
|---|---|---|
| $s^3$ | 1 | 3 |
| $s^2$ | 6 | $C$ |
| $s^1$ | $\dfrac{18 - C}{6}$ | 0 |
| $s^0$ | $C$ | |

To be stable, all the entries in the first column must be positive. This requires that $0 < C < 18$. 11 is the only option that ensures system stability.

*The answer is (C).*

**60.** The thermal agitation noise, or Johnson noise, is

$$V_{\text{noise}} = \sqrt{4\kappa T R \Delta f}$$

The term $\kappa$ is Boltzmann's constant, with a value of $1.3807 \times 10^{-23}$ J/K. The term $T$ is the absolute temperature, $R$ is the resistance, and the bandwidth is represented by $\Delta f$. Substitute the given information, noting that 25°C is equal to 298K, to find the thermal agitation noise.

$$\begin{aligned} V_{\text{noise}} &= \sqrt{4\kappa T R \Delta f} \\ &= \sqrt{\begin{aligned} &(4)\left(1.3807 \times 10^{-23} \frac{\text{J}}{\text{K}}\right)(298\text{K}) \\ &\times (20 \times 10^3 \text{ }\Omega)(1000 \text{ Hz}) \end{aligned}} \\ &= 5.74 \times 10^{-7} \text{ V} \quad (0.6 \text{ } \mu\text{V}) \end{aligned}$$

*The answer is (D).*

**61.** The curl of a given vector, $\mathbf{F}$, can be found from determinant mathematics using the following general formula.

$$\text{curl } \mathbf{F} = \nabla \times \mathbf{F} = \begin{vmatrix} \mathbf{i} & \mathbf{j} & \mathbf{k} \\ \dfrac{\partial}{\partial x} & \dfrac{\partial}{\partial y} & \dfrac{\partial}{\partial z} \\ F_x & F_y & F_z \end{vmatrix}$$

$$= \left( \frac{\partial F_z}{\partial y} - \frac{\partial F_y}{\partial z} \right) \mathbf{i} + \left( \frac{\partial F_x}{\partial z} - \frac{\partial F_z}{\partial x} \right) \mathbf{j}$$

$$+ \left( \frac{\partial F_y}{\partial x} - \frac{\partial F_x}{\partial y} \right) \mathbf{k}$$

Substitute the given values for the magnetic field strength and apply the formula. Units are not shown, as is common practice within a matrix. (If the units of $\mathbf{H}$ are A/m, the units of $\mathbf{J}$ will be A/m$^2$.)

$$\mathbf{J} = \text{curl } \mathbf{H} = \begin{vmatrix} \mathbf{i} & \mathbf{j} & \mathbf{k} \\ \dfrac{\partial}{\partial x} & \dfrac{\partial}{\partial y} & \dfrac{\partial}{\partial z} \\ (5x+3y) & (-5y-3z) & (5z-3x) \end{vmatrix}$$

$$= \big( 0 - (-3) \big) \mathbf{i} + \big( 0 - (-3) \big) \mathbf{j} + (0 - 3) \mathbf{k}$$

$$= 3\mathbf{i} + 3\mathbf{j} - 3\mathbf{k} \text{ A/m}^2$$

*The answer is (A).*

**62.** The output power can be found algebraically using the concept of decibels. First, express the input power in dBm.

$$P_{\text{in,dBm}} = 10 \log \frac{P_{\text{out}}}{P_{\text{ref}}}$$

$$= 10 \log \frac{30 \text{ mW}}{1 \text{ mW}}$$

$$= 14.77 \text{ dBm}$$

This input power is reduced by the attenuator.

$$P_{c,\text{in,dBm}} = P_{\text{in,dBm}} - P_{\text{loss,dB}}$$

$$= 14.77 \text{ dBm} - 10 \text{ dB}$$

$$= 4.77 \text{ dBm}$$

The values with the ratios dBm and dB can be subtracted, since conversion of dB to dBm involves a factor of $10^{-3}$ in both the numerator and denominator. (See the calculation for $P_{\text{in}}$.)

The coupler splits the power. The power available at the meter is 3 dB less than the power at the coupler input, so

$$P_m = P_{c,\text{in,dBm}} - 3 \text{ dB}$$

$$= 4.77 \text{ dBm} - 3 \text{ dB}$$

$$= 1.77 \text{ dBm}$$

Convert to the unit requested.

$$P_{\text{in,dBm}} = 10 \log \frac{P_{\text{in,W}}}{P_{\text{ref}}}$$

$$P_{\text{in,W}} = \left( \text{antilog} \frac{P_{\text{in,dBm}}}{10} \right) P_{\text{ref}}$$

$$= \left( \text{antilog} \frac{1.77 \text{ dBm}}{10} \right) (1 \text{ mW})$$

$$= 1.50 \text{ mW}$$

The term dBm is unitless. It is shown as a reminder that the reference used for the decibel was milliwatts.

*The answer is (A).*

**63.** The thermal agitation noise is

$$V_{\text{noise}} = \sqrt{4\kappa \, TR(\text{BW})}$$

Convert the ambient temperature of 25°C to the absolute scale.

$$T_{\mathbf{K}} = T_{\circ\text{C}} + 273.15°$$

$$= 25°\text{C} + 273.15°$$

$$= 298.15\text{K}$$

Substitute into the noise formula, noting that $\kappa$ is Boltzmann's constant.

$$V_{\text{noise}} = \sqrt{4\kappa \, TR(\text{BW})}$$

$$= \sqrt{ \begin{array}{c} (4)\left( 1.3805 \times 10^{-23} \dfrac{\text{J}}{\text{K}} \right)(298.15\text{K}) \\ \times (10 \times 10^3 \text{ } \Omega)(100 \times 10^3 \text{ Hz}) \end{array} }$$

$$= 4.06 \times 10^{-6} \text{ V} \quad (4 \text{ } \mu\text{V})$$

*The answer is (C).*

**64.** The op amp responds to the instantaneous values of the voltage. The d'Arsonval meter movement responds to the root-mean-square voltage. The peak voltage associated with the maximum rms reading of 200 V is

$$|V_{\text{peak}}| = \sqrt{2} \, V_{\text{rms}}$$

$$= \sqrt{2} \, (200 \text{ V})$$

$$= 282.8 \text{ V}$$

The op amp maintains linear operation within the range $\mp 13$ V. (The supply voltage is $\pm 16$ V. Nominally, linear operation is maintained when the output voltage is at least 3 V from said voltages.) The op amp peak input voltage of 13 V (both positive and negative) should correspond to 282.8 V. The voltage divider attached to the positive terminal accomplishes this if the divider resistance is

$$(282.8 \text{ V})\left(\frac{R_{\text{div}}}{R_{\text{div}} + 1.5 \times 10^6 \ \Omega}\right) = 13 \text{ V}$$

$$(282.8 \text{ V})R_{\text{div}} = (13 \text{ V})(R_{\text{div}} + 1.5 \times 10^6 \ \Omega)$$

$$(282.8 \text{ V})R_{\text{div}} = (13 \text{ V})R_{\text{div}} + 19.50 \times 10^6 \text{ V·}\Omega$$

$$(282.8 \text{ V})R_{\text{div}} - (13 \text{ V})R_{\text{div}} = 19.50 \times 10^6 \text{ V·}\Omega$$

$$(269.8 \text{ V})R_{\text{div}} = 19.50 \times 10^6 \text{ V·}\Omega$$

$$R_{\text{div}} = \frac{19.50 \times 10^6 \text{ V·}\Omega}{269.8 \text{ V}}$$
$$= 7.23 \times 10^4 \ \Omega \quad (0.07 \text{ M}\Omega)$$

Though the answer is rounded, high-precision resistors are used for scaling these types of op amp circuits.

*The answer is (B).*

**65.** The transistor is in saturation with the base-emitter and base-collector junctions forward biased. The first-order model is

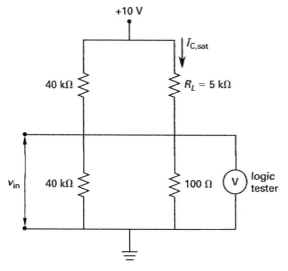

The current through the load resistor is the saturation current.

$$I_{C,\text{sat}} \approx \frac{V_{\text{CC}}}{R_L} = \frac{10 \text{ V}}{5 \times 10^3 \ \Omega}$$
$$= 2 \times 10^{-3} \text{ A} \quad (2 \text{ mA})$$

The output voltage is low (near 0 V) when the transistor is saturated. The manufacturer's data sheet refers to this as "logic 1," so a low voltage is considered to be logic 1 (i.e., TRUE). The logic is "negative."

*The answer is (C).*

**66.** The 3 dB downpoint corresponds to the cutoff frequency as shown.

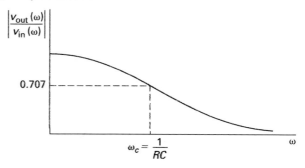

The cutoff frequency for the RC circuit given is

$$\omega_c = 2\pi f = \frac{1}{RC}$$

Substitute the given quantities and solve for the capacitance.

$$2\pi f = \frac{1}{RC}$$
$$C = \frac{1}{2\pi fR} = \frac{1}{2\pi(60 \text{ Hz})(50 \times 10^3 \ \Omega)}$$
$$= 5.31 \times 10^{-8} \text{ F} \quad (0.05 \ \mu\text{F})$$

Capacitor $C_2$ thus meets the stated requirement.

*The answer is (B).*

**67.** The circuit, a difference amplifier, is redrawn with parameters defined as

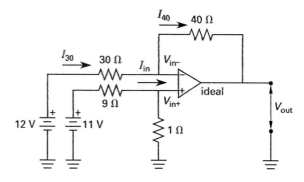

Using $V_1$ and the concept of a voltage divider, the voltage at the positive input is

$$V_{\text{in}^+} = V_1\left(\frac{R_1}{R_1 + R_9}\right)$$
$$= (11 \text{ V})\left(\frac{1 \text{ }\Omega}{1 \text{ }\Omega + 9 \text{ }\Omega}\right)$$
$$= 1.1 \text{ V}$$

Given that ideal conditions may be applied, a virtual short circuit exists between the positive and negative inputs. That is, the voltages are identical (since the current through the feedback resistor, $I_{40}$, drives the value of $V_{\text{in}^-}$ to the value of $V_{\text{in}^+}$). Thus,

$$V_{\text{in}^-} = V_{\text{in}^+}$$
$$= 1.1 \text{ V}$$

Apply Ohm's law to find the current through the 30 $\Omega$ resistor.

$$I_{30} = \frac{V_2 - V_{\text{in}^-}}{R_{30}}$$
$$= \frac{12 \text{ V} - 1.1 \text{ V}}{30 \text{ }\Omega}$$
$$= 0.363 \text{ A}$$

Again, since the conditions are ideal, the input impedance is infinite and the input current is zero. Thus,

$$I_{\text{in}^-} = 0 \text{ A}$$
$$I_{30} = I_{40}$$
$$= 0.363 \text{ A}$$

Use Kirchhoff's voltage law in the feedback loop to find the output voltage.

$$V_{\text{out}} = V_{\text{in}^-} - I_{40}R_{40}$$
$$= 1.1 \text{ V} - (0.363 \text{ A})(40 \text{ }\Omega)$$
$$= -13.420 \text{ V} \quad (-13 \text{ V})$$

*The answer is (B).*

**68.** Network-powered broadband communications systems are covered in NEC Art. 830. As found in Art. 830.100(A)(1), the grounding conductor must be insulated. Per Art. 830.100(A)(3), the minimum size is AWG 14.

*The answer is (B).*

**69.** The constant voltage drop model allows the real diode to be replaced with an ideal diode and a battery source. Silicon diodes are modeled with a 0.7 V source as shown.

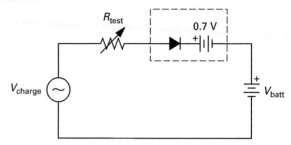

Find the rms value of the source voltage from the peak voltage given, noting that the half-wave rectification means the rms value is one-half of the peak value. (The peak value would be used if reading such values would damage components.)

$$V_{\text{charge,rms}} = \frac{1}{2}V_{\text{charge,peak}}$$
$$= \left(\frac{1}{2}\right)(170 \text{ V})$$
$$= 85 \text{ V}$$

Write Kirchhoff's voltage law around the circuit, starting at the charger voltage source. Include the requirement that the charge voltage be no greater than 5 V above the battery voltage.

$$V_{\text{charge,rms}} - I_{\text{charge}}R_{\text{test}}$$
$$-0.7 \text{ V} - (V_{\text{batt}} + 5 \text{ V}) = 0 \text{ V}$$

Solve for the necessary resistance. Include the requirement for a maximum charge current of 4 $A_{\text{rms}}$.

$$I_{\text{charge,rms}}R_{\text{test}} = V_{\text{charge,rms}} - 0.7 \text{ V} - V_{\text{batt}} - 5 \text{ V}$$
$$R_{\text{test}} = \frac{V_{\text{charge,rms}} - 0.7 \text{ V} - V_{\text{batt}} - 5 \text{ V}}{I_{\text{charge,rms}}}$$
$$= \frac{85 \text{ V} - 0.7 \text{ V} - 30 \text{ V} - 5 \text{ V}}{4 \text{ A}}$$
$$= 12.3 \text{ }\Omega$$

The voltage drop of the diode is insignificant to the overall result. The diode could have been treated as ideal, with satisfactory results.

The resistor must be at least 12.3 $\Omega$ for the voltage requirement between the charge voltage and the battery voltage to be met. 13 $\Omega$ is the minimum value that is required.

*The answer is (B).*

**70.** The radio broadcast regulations are contained in the Code of Federal Regulations. (See Title 47, Chap. 73.14 (47CFR73.14), for an example of such regulations, which in this case sets the bandwidth allowed to AM stations.)

An AM receiver block diagram follows.

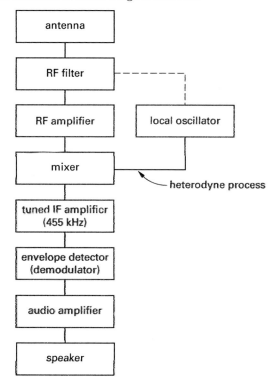

The IF amplifier is tuned to 455 kHz. This is the intermediate frequency (IF) used in the heterodyne process. Tuning to an AM station at 680 kHz requires that the local oscillator be set for

$$
\begin{aligned}
f_{\text{osc}} &= f_{\text{station}} + f_{\text{IF}} \\
&= 680 \text{ kHz} + 455 \text{ kHz} \\
&= 1135 \text{ kHz}
\end{aligned}
$$

Only the higher of the two frequencies generated in the heterodyne process is used. That is, the heterodyne frequency 225 kHz is not used, since to do so would require the oscillator to tune to a wider band, proportionally.

*The answer is (D).*

**71.** The power density of the signal is

$$
\begin{aligned}
p_r &= \left( \frac{D_T}{4\pi r^2} \right) P_T \\
&= \left( \frac{1.4}{4\pi (5 \times 10^3 \text{ m})^2} \right) (5 \times 10^3 \text{ W}) \\
&= 22.28 \times 10^{-6} \text{ W/m}^2 \quad (22 \ \mu\text{W/m}^2)
\end{aligned}
$$

*The answer is (A).*

**72.** The number of electrons passing through the breaker can be found using unit analysis.

$$
\dot{Q}_{\text{min}} = \frac{\text{current rating}}{\text{charge per electron}}
$$

$$
= \frac{(15 \text{ A}) \left( \dfrac{1 \ \frac{\text{C}}{\text{s}}}{1 \text{ A}} \right) \left( 60 \ \dfrac{\text{s}}{\text{min}} \right)}{1.602 \times 10^{-19} \ \dfrac{\text{C}}{\text{electron}}}
$$

$$
= 5.62 \times 10^{21} \text{ electrons/min}
$$

The percentage of free electrons per minute as compared to free electrons (per meter of wire) is

$$
\begin{aligned}
N_\% &= \frac{\left( 5.62 \times 10^{21} \ \dfrac{\text{electrons}}{\text{min}} \right)(1 \text{ min})}{2.77 \times 10^{23} \text{ free electrons}} \times 100\% \\
&= 2.03\% \quad (2\%)
\end{aligned}
$$

Since the typical household current is AC, a single free electron may be accounted for numerous times in the 2.03%.

*The answer is (C).*

**73.** The total energy, or work, from time $t = 0$ s onward is

$$
W = \int_0^\infty vi\, dt
$$

$$
= \int_0^\infty \left((25\ \text{V})(1 - e^{-3000t})\right)\left((5\ \text{A})e^{-3000t}\right) dt
$$

$$
= \int_0^\infty \left(25\ \text{V} - (25\ \text{V})e^{-3000t}\right)\left((5\ \text{A})e^{-3000t}\right) dt
$$

$$
= \int_0^\infty \left((125\ \text{W})e^{-3000t} - (125\ \text{W})e^{-6000t}\right) dt
$$

$$
= 125\ \text{W} \int_0^\infty e^{-3000t}\, dt - 125\ \text{W} \int_0^\infty e^{-6000t}\, dt
$$

$$
= (125\ \text{W})\left(\frac{e^{-3000t}}{-3000}\right)\Bigg|_0^\infty - (125\ \text{W})\left(\frac{e^{-6000t}}{-6000}\right)\Bigg|_0^\infty
$$

$$
= (125\ \text{W})\left(0\ \text{s} - \left(\frac{1}{-3000}\ \text{s}\right)\right) - (125\ \text{W})
$$

$$
\times \left(0\ \text{s} - \left(\frac{1}{-6000}\ \text{s}\right)\right)
$$

$$
= \frac{125}{3000}\ \text{J} - \frac{125}{6000}\ \text{J}
$$

$$
= 2.083 \times 10^{-2}\ \text{J} \quad (20\ \text{mJ})
$$

*The answer is (B).*

**74.** The steady-state value of the total charge on the capacitor is

$$
q = Cv
$$

$$
= (50 \times 10^{-6}\ \text{F})
$$

$$
\times (169.7\ \text{V})\sin\left((377\ \text{Hz})(6\pi \times 10^{-3}\ \text{s})\right)
$$

$$
= (50 \times 10^{-6}\ \text{F})(169.7\ \text{V})\sin(7.106\ \text{rad})
$$

$$
= (50 \times 10^{-6}\ \text{F})(169.7\ \text{V})(0.733)
$$

$$
= 6.2 \times 10^{-3}\ \text{C}
$$

*The answer is (C).*

**75.** The current in the capacitor at 0.625 ms is approximated by

$$
i(t) = C\frac{dv(t)}{dt}
$$

$$
= C\left(\frac{d(339.5\ \text{V})\sin 2512t}{dt}\right)
$$

$$
= C(339.5\ \text{V})\left(\frac{d(\sin 2512t)}{dt}\right)
$$

$$
= C(339.5\ \text{V})(2512\ \text{Hz})\cos 2512t
$$

$$
= (30 \times 10^{-6}\ \text{F})(339.5\ \text{V})(2512\ \text{Hz})
$$

$$
\times \cos\left((2512\ \text{Hz})(0.625 \times 10^{-3}\ \text{s})\right)
$$

$$
= 0.020\ \text{A}
$$

*The answer is (C).*

**76.** The force on charge $Q_2$ is

$$
\mathbf{F_2} = \left(\frac{Q_2 Q_1}{4\pi\epsilon_0 r_{12}^2}\right)\mathbf{a_{12}} \tag{I}
$$

The unknowns are the value of the unit vector, $\mathbf{a_{12}}$, and the value of $r_{12}$. The vector $\mathbf{r_{12}}$ is determined from

$$
\mathbf{r_{12}} = (x_2 - x_1)\mathbf{i} + (y_2 - y_1)\mathbf{j} + (z_2 - z_1)\mathbf{k}
$$

$$
= (0\ \text{m} - 2\ \text{m})\mathbf{i} + (0\ \text{m} - 0\ \text{m})\mathbf{j}
$$

$$
+ (3\ \text{m} - 0\ \text{m})\mathbf{k}
$$

$$
= (-2\ \text{m})\mathbf{i} + (3\ \text{m})\mathbf{k}
$$

The magnitude of $\mathbf{r_{12}}$ is the value $r_{12}$.

$$
r_{12} = |\mathbf{r_{12}}| = \sqrt{(-2\ \text{m})^2 + (3\ \text{m})^2}
$$

$$
= \sqrt{13}\ \text{m} \tag{II}
$$

The unit vector $\mathbf{a_{12}}$ is

$$
\mathbf{a_{12}} = \frac{\mathbf{r_{12}}}{r_{12}} = \frac{(-2\ \text{m})\mathbf{i} + (3\ \text{m})\mathbf{k}}{\sqrt{13}\ \text{m}}
$$

$$
= -0.5547\mathbf{i} + 0.8321\mathbf{k} \tag{III}
$$

Substitute the given information, Eqs. II and III, into Eq. I.

$$\mathbf{F}_2 = \left(\frac{Q_2 Q_1}{4\pi\epsilon_0 r_{12}^2}\right)\mathbf{a}_{12}$$

$$= \left(\frac{(20\times 10^{-6}\ \mathrm{C})(20\times 10^{-6}\ \mathrm{C})}{4\pi\left(8.854\times 10^{-12}\ \dfrac{\mathrm{F}}{\mathrm{m}}\right)(\sqrt{13}\ \mathrm{m})^2}\right)$$

$$\times(-0.5547\mathbf{i} + 0.8321\mathbf{k})$$

$$= (0.2765\ \mathrm{N})(-0.5547\mathbf{i} + 0.8321\mathbf{k})$$

$$= (-0.1534\ \mathrm{N})\mathbf{i} + (0.2301\ \mathrm{N})\mathbf{k}$$

$$\big((-0.2\ \mathrm{N})\mathbf{i} + (0.2\ \mathrm{N})\mathbf{k}\big)$$

The resultant force on $Q_2$, with a magnitude of approximately 0.1 N, is shown.

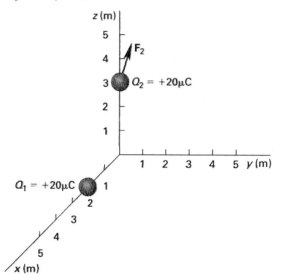

**The answer is (A).**

**77.** The magnitude of an electric field is

$$E = |\mathbf{E}| = \frac{Q}{4\pi\epsilon_0 r^2}$$

$$= \frac{1\ \mathrm{C}}{4\pi\left(8.854\times 10^{-12}\ \dfrac{\mathrm{F}}{\mathrm{m}}\right)(10\ \mathrm{m})^2}$$

$$= 8.99\times 10^7\ \mathrm{V/m} \quad (90\ \mathrm{MV/m})$$

**The answer is (C).**

**78.** Using the right-hand rule, placing the thumb in the direction of the current flow, the fingers curl inward toward the direction of the magnetic field. Option D illustrates the result.

**The answer is (D).**

**79.** The Karnaugh map one-squares can be grouped as follows, allowing the "don't care" condition in $m_3$ to take on the value of 1.

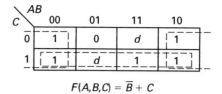

$$F(A,B,C) = \overline{B} + C$$

The function $F(A, B, C)$ is realized with the PLA given by option D.

**The answer is (D).**

**80.** The sum-of-products (SOP) form of the output function uses the minterms of the function. Minterms are those terms in which the value of the function is 1, or true. (Maxterms take on the value of 0.) The minterms, $m$, and maxterms, $M$, for the given function are shown.

| $X$ | $Y$ | $Z$ | $F(X, Y, Z)$ | minterms, $m$ maxterms, $M$ |
|---|---|---|---|---|
| 0 | 0 | 0 | 0 | $M_0$ |
| 0 | 0 | 1 | 1 | $m_1$ |
| 0 | 1 | 0 | 0 | $M_2$ |
| 0 | 1 | 1 | 1 | $m_3$ |
| 1 | 0 | 0 | 0 | $M_4$ |
| 1 | 0 | 1 | 1 | $m_5$ |
| 1 | 1 | 0 | 0 | $M_6$ |
| 1 | 1 | 1 | 1 | $m_7$ |

Minterms exist at input values of one, three, five, and seven. The function, in canonical SOP form, is

$$F(X, Y, Z) = m_1 + m_3 + m_5 + m_7$$

$$= \overline{X}\ \overline{Y}Z + \overline{X}YZ + X\overline{Y}Z + XYZ$$

**The answer is (A).**